Just-in-Time
ALGEBRA

FOR STUDENTS OF CALCULUS IN MANAGEMENT AND THE LIFE SCIENCES

Ronald I. Brent
Guntram Mueller
University of Massachusetts-Lowell

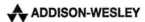 **ADDISON-WESLEY**

An imprint of Addison Wesley Longman, Inc.

Reading, Massachusetts • Menlo Park, California • New York • Harlow, England
Don Mills, Ontario • Sydney • Mexico City • Madrid • Amsterdam

Cover illustration by Toni St. Regis.

Reproduced by Addison-Wesley from camera-ready copy supplied by the author.

ISBN: 0-201-74611-5

3 4 5 6 7 8 9 10 CRS 03 02

Table of Contents

To the Student

Well, here you are in college, taking mathematics, and even calculus. It may have surprised you that you need calculus, since you are not in engineering or the physical sciences. But now that computers are found everywhere, a lot of fields like business, the social sciences, and the life sciences are seeing a more quantitative approach, which often brings in calculus. Whether you are dealing with pricing issues involving elasticity of demand, or safety issues involving maximum drug concentrations in the blood, or even implications of the rate of change of living standards – the ideas and often the methods of calculus have become tools of the trade.

Calculus is built up on algebra and the idea of functions and their graphs. This means (and here you may reflect on your own high school algebra experience) that you could decide that you need to brush up a little on your algebra skills. Well, there is some good news: You don't need all of algebra at once. In fact, you need different parts of algebra at different times during your calculus course. The idea of this book is that the required algebra is built up exactly in the order that you will need to know it in your calculus course. That's why this book is called *Just-in-Time Algebra*. This means that you can focus on the material you need to know, and can spend less time chasing around for it. As for the time you save . . . well, you can probably figure out one or two other things that you'd rather be spending your time on.

What is the best way to use this book? Here's how.

a) At any point in your calculus course: Read the corresponding part of this book as directed by the Table of Contents, having pencil and paper at hand. If there are things you don't understand, or that just seem wrong (it happens!), put a question mark in the book and ask your professor in class. Make sure it all makes sense <u>to you</u>. <u>Your mind is the ultimate judge!</u>

b) It's best if you do <u>all</u> of the exercises. The exercise sets are kept fairly short, and each exercise brings out one or more new features.

A word about calculators: They are a great way to carry out obnoxious calculations easily and accurately, as well as a great tool for exploring different possibilities. An obvious limitation is that you can make mistakes in using them, and therefore you must always use a separate, independent way of carrying out the calculations to make sure your answer isn't totally "off the wall." Sometimes you can use rough approximations to do the calculations with pencil and paper, or even do them entirely in your head. The point we're getting at is: Don't use your calculator to "fly blind," or you can expect to have some nasty crackups.

Calculus isn't the easiest thing on earth, but neither is it the toughest. You <u>can</u> succeed in calculus, by making sure you understand what's in the book and the classes, and by keeping up-to-date doing the homework exercises. You know how important practicing is in sports, and in music. It's very much like that also in calculus.

To the Instructor

What calculus instructors have not, at one time or another, wrung their hands or pulled out their hair in despair over the level of preparedness of their students? This book is an attempt to try to salvage whatever is left of your hair.

As you can see from the Table of Contents, this book is organized by topic from most standard calculus courses for management and the life sciences. When limits are to be calculated, for example, the student needs factoring skills, so here are the factoring methods, all lined up at just the time the student needs them. There is too little room in calculus books for this material on factoring, and sending students to the library to look up factoring methods on their own does not produce the desired results.

Similarly, when studying the idea of the derivative, here is a review of rational operations on rational expressions. Doing implicit differentiation? Here's how to solve equations of degree 1, even strange-looking ones. The idea is that the student does not need to hunt through libraries and unfamiliar algebra texts to look for what he or she might need. Rather, it is all lined up here, "just-in-time," exactly when needed.

What is not in this book? This book does not contain material that is presented in standard calculus texts. So, for example, limits are not included, since they are covered in the standard texts. The idea is to keep this book light enough for students to actually carry it and use it – and cheap enough for them to afford it.

This book is intended to be used in one of two ways:

a) As a second text for calculus courses whose syllabi officially include time for a review of algebra. The organization of this book is a key feature: It can be used with almost all standard calculus texts, including the reform texts, in a manner that interleaves the topics from algebra with those from calculus. By timing certain topics in algebra to be given just before, or while, they are needed in calculus, we address the often-heard student complaint of irrelevance. For example, factoring is made more pertinent by tying it in with limit problems that can't be done except by factoring. This will enhance the motivation to learn factoring, a topic that can well use all the motivation it can get.

b) As a companion to a standard calculus course, for those students who in fact need a little well-timed help in algebra. This book is written in an easy style that can be understood by all students on their own, without input by the instructor. It is there as a reference, a guide, a handbook, a companion on a journey that is rewarding, but where the traveler could sometimes use a little support.

The idea for both uses is the same: the deft timing of the algebra review, topic-by-topic, just at the time that it is needed in calculus. It makes for a more relevant presentation of the review topic, which allows the student to be more interested in it because he or she is about to have to use it in calculus homework, that night or the next.

Acknowledgments

We wish to acknowledge all those who helped in this venture, from our wives, Leor and Edie, and our children, Sarah and Adam, and Ariadne, to the many students who have helped test the book. We especially wish to thank Laurie Rosatone, Jennifer Wall, and Rachel St. Pierre, at Addison Wesley Longman, and all the people who reviewed the manuscript. They provided us with excellent insights and many valuable suggestions.

Once in a while you get shown the light, in the strangest of places, if you look at it right.

– Robert Hunter

University of Massachusetts Lowell

Ronald I. Brent

Guntram Mueller

Chapter 1

Numbers and Their Disguises

Every number can be written in many different forms. For example, the numbers $\frac{8}{12}$, $\frac{10}{15}$, $\frac{3}{4 + \pi/2\pi}$, and even $\frac{|4x| + 4}{|6x| + 6}$ are all really just different ways of writing the number $\frac{2}{3}$. (Check it out; don't take <u>our</u> word for it.) For most purposes, the idea is to keep things as simple as possible, and just to use the form $\frac{2}{3}$.

Sometimes it is better to have a mathematical expression written as a product, other times it is better to have a sum; it all depends on what you need to do with it. In any event, <u>changing the form of an expression is something you have to do all the time</u>. Correctly! It's the nuts and bolts of mathematics, and is used in all the sciences and engineering, and even in economics and medicine.

1.1 <u>Brackets</u>

Brackets are ways of "packaging" or grouping numbers together.

Example 1:

$$5 - (1 + \frac{1}{2}) + (3 - 4) - (7 - \frac{1}{2})$$
$$= 5 - (1\tfrac{1}{2}) + (-1) - (6\tfrac{1}{2})$$
$$= 5 - \frac{3}{2} - 1 - \frac{13}{2}$$
$$= 5 - 1 - \frac{3}{2} - \frac{13}{2}$$
$$= 4 - \frac{16}{2} = 4 - 8 = -4 \quad \blacksquare$$

You may prefer to get rid of the brackets. Here's how: a) if there's a "+" in front, leave all the signs of the terms inside as they are; b) if there is a "–" in front, change all the signs of the terms inside.

Example 2:

$$5 - \left(1 + \frac{1}{2}\right) + (3 - 4) - \left(7 - \frac{1}{2}\right)$$

$$= 5 - 1 - \frac{1}{2} + 3 - 4 - 7 + \frac{1}{2}$$

$$= 5 - 1 + 3 - 4 - 7 - \frac{1}{2} + \frac{1}{2}$$

$$= -4 \quad \blacksquare$$

Example 3:   Simplify $3 - 2(8 + 1) - 3(5 - 7)$.

Solution: Method 1: $3 - 2(9) - 3(-2) \ = \ 3 - 18 + 6 \ = \ -9$

Method 2: $3 - 16 - 2 \ - 15 + 21 \ = \ -9 \quad \blacksquare$

Notice that in the second method in Example 3, the 8 and 1 were both multiplied by -2, and the 5 and -7 were multiplied by -3. This is based on the law of distributivity – namely, $a(b + c) = ab + ac$. Method 1 is easier in this case, but in many other cases method 2 will be needed. <u>You've got to know both</u>!

In algebra, letters stand for numbers, so the laws of arithmetic apply to them in exactly the same way.

Example 4:

$$xy - (2x - y) - 2y(1 - x)$$

$$= xy - 2x + y - 2y + 2yx$$

$$= 3xy - 2x - y \quad \blacksquare$$

Notice that xy and $2yx$ add up to $3xy$.

Example 5:

$$3x^2y - (x^2 - y^3) - 2y(x - y)$$

$$= 3x^2y - x^2 + y^3 - 2xy + 2y^2 \quad \blacksquare$$

Notice the "+" in front of the y in Example 4 and the y^3 in Example 5, as well as the use of the distributive law in both examples. If the exponents in the last example are confusing you, skip ahead to Section 1.4 and then return here.

Exercises 1.1   Simplify:

1) $4 - (5 - 3) + 3(4 - 7) - 4(1 + 2)$

2) $-2(3 - 4) + 4(5 + 3) - 3(2 - 6)$

3) $4xy - (x - 2xy) - 2y(x - 1)$

4) $(s - t) - (u - t) - (v - u) - (s - v)$

5) $2x(y - 3) - y(x + xy) + 2y(x + 1)$

6) $x(y + z) - z(x + y) + 2y(x - z) - x(3y - 2z)$

7) $xy(x + y^2) - (2x^2y^2 - 2xy^3) - 2y^2(x^2 - y + xy)$

8) $xy^2(x^2 + y^2) - 3x(2xy^2 - 2xy^3) + 2y^2(x - xy^2 - x^2)$

9)   Consider simplifying the expression $-2(7 - 4) + 5(2 + 3) - 2(2 - 6)$ by first computing the expression within the brackets and then by using the distributive law. Keep track of the total number of computations (additions and subtractions or multiplications) in each method. Which way is "cheaper?"

1.2 **Multiplying and Dividing Fractions** (We'll do adding and subtracting later.)

Multiplying fractions is the easiest manipulative task. The numerator is the product of the given numerators, and the denominator is the product of the given denominators. That is

$$\frac{a}{b} \cdot \frac{c}{d} = \frac{a \cdot c}{b \cdot d}.$$

We use the symbol "·" to denote multiplication, although sometimes we will omit the symbol altogether.

Example 1:

a) $\dfrac{2}{3} \cdot \dfrac{5}{7} = \dfrac{2 \cdot 5}{3 \cdot 7} = \dfrac{10}{21}$

b) $\dfrac{1}{9} \cdot \left(-\dfrac{5}{8}\right) = \dfrac{1}{9} \cdot \dfrac{-5}{8} = \dfrac{1 \cdot (-5)}{9 \cdot 8} = \dfrac{-5}{72}$ ∎

This rule can be extended to multiplying more than two fractions simply by multiplying across all the numerators and denominators.

Example 2:

a) $\dfrac{1}{4} \cdot \dfrac{7}{5} \cdot \dfrac{3}{8} = \dfrac{1 \cdot 7 \cdot 3}{4 \cdot 5 \cdot 8} = \dfrac{21}{160}$

b) $\dfrac{-1}{7} \cdot \dfrac{3}{8} \cdot \dfrac{-2}{\pi} = \dfrac{(-1)(3)(-2)}{7 \cdot 8 \cdot \pi} = \dfrac{6}{56\pi} = \dfrac{3}{28\pi}$ ∎

Of course, multiplying a number by 1 produces the same number, so: $\dfrac{5}{6} \cdot 1 = \dfrac{5}{6} \cdot \dfrac{4}{4} = \dfrac{20}{24}$. Going backward is usually more important: $\dfrac{20}{24} = \dfrac{5 \cdot 4}{6 \cdot 4} = \dfrac{5 \cdot \cancel{4}}{6 \cdot \cancel{4}} = \dfrac{5}{6}$. Here's the key point: you can cancel the factor 4 in the numerator and the denominator. More generally, if a number $c \neq 0$ is a factor of both the top and bottom of a fraction, it may be canceled. When all those common factors are canceled, the fraction is said to be in lowest terms.

Example 3:   Put $\dfrac{30}{84}$ into lowest terms.

Solution: First factor the numerator and denominator as much as possible and then cancel all common factors.

$$\frac{30}{84} = \frac{\cancel{2}\cdot\cancel{3}\cdot 5}{2\cdot\cancel{2}\cdot\cancel{3}\cdot 7} = \frac{5}{2\cdot 7} = \frac{5}{14} \quad\blacksquare$$

Warning: <u>Make sure you cancel only those numbers that are factors of the entire top and the entire bottom.</u> Don't be tempted to try "creative canceling." Consider the expression

$$\frac{3(8) + 7(5)}{3(4)}.$$

Can you cancel the 3's ? NO NO NO NO NO NO NO NO NO !!!! Get the picture? The problem is that 3 is not a factor of the <u>entire</u> numerator, only of its first term. Hence you cannot simply cancel the 3's. If the expression had been

$$\frac{3(8) + 3(5)}{3(4)},$$

then you would be able to cancel the 3's. In this case, <u>each term of the numerator</u> contains the factor of 3 that is also in the denominator.

$$\frac{3(8) + 3(5)}{3(4)} = \frac{3(8 + 5)}{3(4)} = \frac{\cancel{3}}{\cancel{3}}\cdot\frac{(8+5)}{(4)} = 1\cdot\frac{(8+5)}{(4)} = \frac{13}{4}$$

Dividing by a fraction is done by inverting <u>that</u> fraction and multiplying:

$$\frac{\dfrac{a}{b}}{\dfrac{c}{d}} = \frac{a}{b}\cdot\frac{d}{c} = \frac{a\cdot d}{b\cdot c}.$$

For example,

$$\frac{\dfrac{-1}{3}}{\dfrac{5}{6}} = \frac{-1}{3}\cdot\frac{6}{5} = \frac{-6}{15} = \frac{-2}{5},$$

or you can cancel early:

$$\frac{-1}{1\cancel{0}} \cdot \frac{\cancel{6}^{2}}{5} = \frac{-2}{5}.$$

Example 4: Simplify the following expressions:

$$\text{a)} \quad \frac{\dfrac{2}{3}}{\dfrac{3}{8}} \qquad\qquad \text{b)} \quad \frac{\dfrac{5}{4}}{\dfrac{-10}{3}}$$

Solution:

$$\text{a)} \quad \frac{\dfrac{2}{3}}{\dfrac{3}{8}} \;=\; \frac{2}{3} \cdot \frac{8}{3} \;=\; \frac{16}{9}$$

$$\text{b)} \quad \frac{\dfrac{5}{4}}{\dfrac{-10}{3}} \;=\; \frac{5}{4} \cdot \frac{3}{-10} \;=\; -\frac{3}{8} \quad\blacksquare$$

Example 5: Simplify the expression $\left(\dfrac{x+1}{y}\right) \cdot \left(\dfrac{-y+1}{x}\right)$.

Solution:

$$\left(\frac{x+1}{y}\right) \cdot \left(\frac{-y+1}{x}\right) \;=\; \frac{(x+1)\cdot(-y+1)}{y\cdot x}$$

$$=\; \frac{-xy+x-y+1}{xy}$$

$$=\; \frac{1-xy+x-y}{xy} \quad\blacksquare$$

Example 6: Simplify the expression $\dfrac{\dfrac{x^{2}y}{z}}{\dfrac{xy^{2}}{z^{3}}}$.

Solution:

$$\frac{\dfrac{x^{2}y}{z}}{\dfrac{xy^{2}}{z^{3}}} \;=\; \frac{x^{2}y}{z} \cdot \frac{z^{3}}{xy^{2}}$$

$$=\; \frac{x^{2}yz^{3}}{zxy^{2}} \;=\; \frac{xz^{2}}{y} \quad\blacksquare$$

Again, if you are confused by the exponents, skip ahead to Section 1.4 and then return here.

In calculus, we use numbers called real numbers. These can be thought of as all decimal numbers, including those having an infinite number of digits. The set of real numbers corresponds to the set of points on the number axis. Rational numbers are those real numbers that can be expressed as a quotient of two integers. For example, the numbers $\frac{1}{3}$, $\frac{4}{10}$, -5, $\frac{18}{7}$, $\sqrt{64}$, and 3.857 are all rational numbers. (Convince yourself of this by expressing them as the quotient of two integers.) All other real numbers are called irrational. The number $\sqrt{2}$ can be easily shown to be irrational. The number π is also irrational but that's much harder to prove.

Exercises 1.2   In Exercises 1–12, simplify and reduce to lowest terms. (Some answers may have more than one form.)

1) $\dfrac{5}{16} \cdot \dfrac{8}{10}$

2) $\dfrac{2\pi}{3} \cdot \dfrac{3\pi}{4}$

3) $\dfrac{-1}{3} \cdot \dfrac{-9}{5}$

4) $\dfrac{\frac{4}{75}}{\frac{8}{25}}$

5) $\dfrac{\frac{-7}{51}}{\frac{3}{12}}$

6) $\dfrac{\frac{3\pi}{7}}{\frac{2\pi}{3}}$

7) $\dfrac{7x}{3y} \cdot \dfrac{3y+2}{x}$

8) $\left(\dfrac{x+2}{1+y}\right) \cdot \left(\dfrac{1-y}{x}\right)$

9) $\dfrac{\frac{xy}{w}}{\frac{xy-2x}{w}}$

10) $\dfrac{xy}{wz} \cdot \dfrac{w^2 z}{x^2 y^2}$

11) $\dfrac{\frac{xy}{(x+y)}}{\frac{x^2 y}{(x+y)^3}}$

12) $\dfrac{\frac{xy}{(x-y)}}{\frac{x^2}{y} \cdot \frac{y^3}{x}}$

13)   Draw the number line, and locate the following numbers:

a) $\dfrac{-126}{106}$

b) $-3 + \dfrac{10}{27}$

c) $\dfrac{\pi}{2} - 1$

14)   Express the numbers -5, $\sqrt{64}$, and 3.857 as a quotient of two whole numbers.

1.3 Adding and Subtracting Fractions

Adding and subtracting fractions is easy when the denominators are all the same. For example,

$$\frac{2}{5} + \frac{17}{5} - \frac{4}{5} = \frac{2 + 17 - 4}{5} = \frac{15}{5} = 3.$$

But what if the fractions don't all have the same denominator? Then you first need to rewrite the fractions so that they all have the same denominator, called a <u>common denominator</u>. For example, if you want to add $\frac{2}{3} + \frac{4}{5}$, you can use a common denominator of 15. So:

$$\frac{2}{3} + \frac{4}{5} = \frac{10}{15} + \frac{12}{15} = \frac{22}{15}.$$

Remember, to get a common denominator you can always use the product of the individual denominators ($3 \cdot 5 = 15$), but sometimes even a smaller number will do it. Then, to get the new numerator,

$$\frac{2}{3} = \frac{?}{15},$$

divide 3 into 15, to get 5, then multiply by 2, to get 10. Try it: $\frac{3}{7} = \frac{?}{42}$. (We're doing the opposite of canceling common factors.)

Example 1:   a)   Simplify $\frac{3}{5} + \frac{1}{2} - \frac{2}{3}$.

Solution: $5 \cdot 2 \cdot 3 = 30$ will serve as a common denominator. So:

$$\frac{3}{5} + \frac{1}{2} - \frac{2}{3} = \frac{18}{30} + \frac{15}{30} - \frac{20}{30}$$

$$= \frac{18 + 15 - 20}{30} = \frac{13}{30}$$

b)   Simplify $\frac{1}{6} - \frac{1}{9}$.

Solution: We could use $6 \cdot 9 = 54$ as a common denominator, but even 18 will do nicely because <u>both 6 and 9 divide evenly into 18</u>. So:

$$\frac{1}{6} - \frac{1}{9} = \frac{3}{18} - \frac{2}{18} = \frac{1}{18}$$

c) Simplify $\dfrac{\dfrac{1}{2}+\dfrac{3}{4}}{\dfrac{1}{3}-\dfrac{1}{6}}$.

Solution: $\dfrac{\dfrac{1}{2}+\dfrac{3}{4}}{\dfrac{1}{3}-\dfrac{1}{6}} = \dfrac{\dfrac{2+3}{4}}{\dfrac{2-1}{6}} = \dfrac{\dfrac{5}{4}}{\dfrac{1}{6}} = \dfrac{5}{4}\cdot\dfrac{6}{1} = \dfrac{15}{2}$ ■

You will also have to be comfortable with adding and subtracting fractional expressions involving variables.

Example 2:

$$\dfrac{xy}{z}+\dfrac{x}{5} = \dfrac{5xy}{5z}+\dfrac{xz}{5z} = \dfrac{5xy+xz}{5z} \quad ■$$

Example 3:

$$\dfrac{\dfrac{3a+2b}{5ab}}{\dfrac{a}{b}-\dfrac{b}{a}} = \dfrac{\dfrac{3a+2b}{5ab}}{\dfrac{a^2-b^2}{ab}}$$

$$= \dfrac{3a+2b}{5ab}\cdot\dfrac{ab}{a^2-b^2}$$

$$= \dfrac{3a+2b}{5a^2-5b^2} \quad ■$$

Exercises 1.3   Express as a single fraction and simplify:

1) $\dfrac{1}{3} + \dfrac{1}{4}$

2) $\dfrac{7}{6} + \dfrac{5}{24}$

3) $\dfrac{2}{5} - \dfrac{1}{2} + \dfrac{1}{3}$

4) $\dfrac{1}{2} - \dfrac{1}{4} + \dfrac{1}{8} - \dfrac{1}{16}$

5) $\dfrac{1}{2} + \dfrac{4}{3} - \dfrac{2}{5} - \dfrac{3}{15}$

6) $\dfrac{5}{6} - \dfrac{4}{3} + \dfrac{2}{9} - \dfrac{3}{2}$

7) $\dfrac{\dfrac{1}{3} + \dfrac{2}{5}}{\dfrac{3}{2}}$

8) $\dfrac{\dfrac{1}{4} - \dfrac{2}{3}}{\dfrac{3}{2} - \dfrac{2}{5}}$

9) $\dfrac{\dfrac{2}{7} + \dfrac{1}{3}}{\dfrac{4}{3} + \dfrac{2}{5}} + \dfrac{1}{3}$

10) $\dfrac{1}{x} + \dfrac{1}{y}$

11) $\dfrac{1}{y} - \dfrac{1}{x}$

12) $\dfrac{4}{x} - \dfrac{2}{y} + \dfrac{1}{z}$

13) $\dfrac{1}{x} - \dfrac{x+1}{xy} + \dfrac{x-2}{xz}$

14) $\dfrac{\dfrac{1}{y} - \dfrac{x}{z}}{\dfrac{1}{z} - \dfrac{1}{x}}$

15) $\dfrac{\dfrac{1}{st} - \dfrac{1}{w}}{\dfrac{1}{tw} - \dfrac{2}{s}}$

16) $\dfrac{y}{x} - \dfrac{x}{y}$

17) $\dfrac{4yz}{x^2} - \dfrac{2z}{xy^2} + \dfrac{1}{xyz}$

18) $\dfrac{\dfrac{1}{x} - \dfrac{x}{y}}{\dfrac{2y}{x} + \dfrac{2x}{y}} + \dfrac{x-y}{xyz}$

1.4 <u>Exponents</u>

Positive whole number exponents are a simple mathematical notation to represent repeated multiplication by the same factor. But you can also have negative numbers and 0 as exponents. Consider the table below.

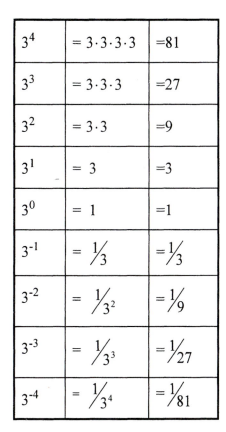

3^4	$= 3 \cdot 3 \cdot 3 \cdot 3$	$= 81$
3^3	$= 3 \cdot 3 \cdot 3$	$= 27$
3^2	$= 3 \cdot 3$	$= 9$
3^1	$= 3$	$= 3$
3^0	$= 1$	$= 1$
3^{-1}	$= \frac{1}{3}$	$= \frac{1}{3}$
3^{-2}	$= \frac{1}{3^2}$	$= \frac{1}{9}$
3^{-3}	$= \frac{1}{3^3}$	$= \frac{1}{27}$
3^{-4}	$= \frac{1}{3^4}$	$= \frac{1}{81}$

Exponents increase by 1. ↑   Exponents decrease by 1. ↓   Numbers are multiplied by 3. ↑   Numbers are divided by 3. ↓

See the pattern? Let's summarize: $3^n = \underbrace{3 \cdot 3 \cdot 3 \cdot \,\cdots\, \cdot 3 \cdot 3}_{n \text{ factors}}$, $3^{-n} = \dfrac{1}{3^n}$, and $3^0 = 1$. In general, given any number x and any positive integer n,

$$x^n = \underbrace{x \cdot x \cdot x \cdot \,\cdots\, \cdot x \cdot x}_{n \text{ factors}} \,,$$

$$x^{-n} = \frac{1}{x^n} \ \text{ for } x \neq 0,$$

and
$$x^0 = 1 \text{ for } x \neq 0.$$

In all of these rules, the number x is called the <u>base</u>, while the number n is called the <u>exponent</u>. Notice that

$$x^m \cdot x^n = x^{m+n},$$

$$x^m / x^n = x^{m-n},$$

and

$$\left(x^m\right)^n = x^{m \cdot n}.$$

These rules are also true for values of m and n that are not integers. We will deal with such exponents later.

Example 1:   a) $\dfrac{4^2 - 1}{3^3 - 2^2} = \dfrac{16 - 1}{27 - 4} = \dfrac{15}{23}$

b) $5^{-1} + 3^{-1} = \dfrac{1}{5} + \dfrac{1}{3} = \dfrac{3 + 5}{15} = \dfrac{8}{15}$

c) $\dfrac{3 \cdot 8^2}{9 \cdot 8^3} = \dfrac{3 \cdot 8^2}{9 \cdot 8^3} = \dfrac{1}{3} \cdot \dfrac{8^2}{8^3} = \dfrac{1}{3} \cdot \dfrac{1}{8} = \dfrac{1}{24}$ ∎

Example 2: $\dfrac{x^2 y^5}{x^{-3}} \div \dfrac{x^{-5} y^4}{x^3}$

$$= \left(\dfrac{x^2}{x^{-3}} y^5\right) \div \left(\dfrac{x^{-5}}{x^3} y^4\right)$$

$$= x^5 y^5 \div x^{-8} y^4$$

$$= \dfrac{x^5 y^5}{x^{-8} y^4} = x^{13} y \quad ∎$$

Example 3: $\dfrac{1}{a^3} - \left(\dfrac{1}{a^5} - \dfrac{1}{a^2}\right)$

$$= \dfrac{1}{a^3} - \dfrac{1}{a^5} + \dfrac{1}{a^2}$$

$$= \dfrac{a^2 - 1 + a^3}{a^5} \quad ∎$$

Example 4: $\left(\dfrac{x^{-2}}{x^8}\right)^{-2}$

$$= \left(x^{-2} x^{-8}\right)^{-2}$$

$$= \left(x^{-10}\right)^{-2} = x^{20} \quad ∎$$

Exercises 1.4   In Exercises 1–9, simplify the expressions.

1) $\dfrac{4^{-1}5^2}{2^2 3^{-2}}$

2) $\dfrac{3^5 2^3}{4^2 3^3}$

3) $\dfrac{5^3}{3^{-1}5^2 + 4^{-1}5^3}$

4) $4^3 \left(\dfrac{1}{4}\right)^2 3^{-4}$

5) $\dfrac{1}{2^{-3}} - \dfrac{1}{2} + \dfrac{1}{5^{-2}}$

6) $x^2 y^{-2} z^3 x^{-2} y^3 z^5$

7) $\left(2^2\right)^{-1}$

8) $\left(x^2\right)^{34}$

9) $\left(\dfrac{1}{x^2}\right)^{34}$

10)   Show by example that $\left(x^{-2} + y^{-2}\right)^2 \neq x^{-4} + y^{-4}$, that is find values for x and y so that the two sides are unequal for those values. (Hint: Just dive in and try some. Maybe you'll be lucky.)

Simplify:

11) $\dfrac{x^{-1}y^2}{y^2 x^{-2}}$

12) $\dfrac{\left(x^2 y^{-3}\right)^2}{\left(y^{-3} x^{-2}\right)^{-2}}$

13) $\dfrac{x^2 y}{x^3} \div \dfrac{x^{-3} y^6}{y^4}$

14) $\dfrac{\dfrac{x^2 y^{-3}}{3z^2} - \dfrac{z^{-3} y^{-3}}{3x^2}}{\dfrac{x^{-4} y^2}{3z^{-2}}}$

15) $\left(x^{-1} + y^{-1}\right)^{-1}$

1.5 <u>Roots</u> (also called radicals)

Definition: We say that the number y is a <u>square root</u> of the number x if $y^2 = x$. A <u>cube root</u> of the number x is a number y such that $y^3 = x$, and so on. In general, if n is any positive integer, we say that the number y is an "nth root" of the number x if $y^n = x$. The number n is called the <u>order</u> of the root.

<u>For roots of even order:</u>

a) The number 16 has two square roots, 4 and -4. The "radical" symbol $\sqrt{}$ means the <u>positive</u> square root always! So, $\sqrt{16} = 4$, but $\sqrt{16} \neq -4$. We say that 16 has two square roots: $\sqrt{16}$ and $-\sqrt{16}$ (that is 4 and -4). The situation is similar for higher even-order roots: Each positive number x has two even-order nth roots, denoted $\pm\sqrt[n]{x}$, where $\sqrt[n]{x}$ is always taken as the <u>positive</u> nth root. For example: $\sqrt[4]{16} = 2$, since $2^4 = 16$; also $\sqrt[6]{64} = 2$, $\sqrt[4]{81} = 3$, and $-\sqrt{81} = -9$.

b) Since no real number multiplied by itself an even number of times can produce a negative number, negative numbers have no real roots of even order. So $\sqrt{-4}$ is not a <u>real</u> number. (Do you know what it is ?) Notice that we usually write $\sqrt{}$ instead of $\sqrt[2]{}$.

To sum up:

  i)   Every positive number x has two real nth roots if n is even – namely, $\sqrt[n]{x}$ and $-\sqrt[n]{x}$.

  ii)   Only positive numbers, and 0, have even-order roots that are real numbers.

<u>For roots of odd order:</u>

Things are a little different with odd-order roots. All numbers have exactly one odd-order nth root, denoted $\sqrt[n]{x}$. The number 8 has exactly 1 real cube root – namely, 2. So, $\sqrt[3]{8} = 2$. But notice $\sqrt[3]{-8} = -2$, because $(-2)^3 = -8$, and $\sqrt[3]{-27} = -3$, because $(-3)^3 = -27$. Get it? Multiplying a negative number by itself an odd number of times produces a negative number!

<u>An alternative notation</u> for $\sqrt[n]{a}$ is $a^{1/n}$. Both notations mean exactly the same thing. For example, $8^{1/3} = 2$, $25^{1/2} = 5$ (not -5), and $(-16)^{1/2}$ is not defined (as a real number).

Definition: We can also define <u>fractional exponents</u>: if $\dfrac{m}{n}$ is in lowest terms, and if n and a are such that $\sqrt[n]{a}$ makes sense, then we define $a^{m/n}$ to be $(\sqrt[n]{a})^m$, which equals $\sqrt[n]{a^m}$. If a is much larger than 1, it is usually easier to take the root first; it keeps the numbers down.

Example 1: a) $\qquad 8^{\frac{2}{3}} = (\sqrt[3]{8})^2 = 2^2 = 4$

b) $\qquad \left(\dfrac{-1}{27}\right)^{\frac{4}{3}} = \left(\sqrt[3]{\dfrac{-1}{27}}\right)^4 = \left(\dfrac{-1}{3}\right)^4 = \dfrac{1}{81}$

c) $\qquad (-32)^{\frac{4}{5}} = (\sqrt[5]{-32})^4 = (-2)^4 = 16$ ∎

Laws of Exponents: Let r and s be any rational numbers. Let a and b be any real numbers. Then, in each of the following, if the expressions on both sides exist, they will be equal. (When might they not exist?)

1. $\qquad a^r \cdot a^s = a^{r+s}$

2. $\qquad \dfrac{a^r}{a^s} = a^{r-s}$

3. $\qquad (a^r)^s = a^{r \cdot s}$

4. $\qquad (ab)^r = a^r b^r$

5. $\qquad \left(\dfrac{a}{b}\right)^r = \dfrac{a^r}{b^r}$

6. $\qquad \left(\dfrac{a}{b}\right)^{-r} = \left(\dfrac{b}{a}\right)^r = \dfrac{b^r}{a^r}$

Remark: These laws also contain the laws of roots. For example, law # 4 for $r = \dfrac{1}{2}$ says: $(ab)^{\frac{1}{2}} = a^{\frac{1}{2}} b^{\frac{1}{2}}$, and so $\sqrt{ab} = \sqrt{a}\,\sqrt{b}$ if $a \geq 0$ and $b \geq 0$. Similarly, by # 5 we have $\sqrt{\dfrac{a}{b}} = \dfrac{\sqrt{a}}{\sqrt{b}}$ if $a \geq 0$ and $b > 0$. Is something similar true for addition? See the exercises below.

Exercises 1.5   In Exercises 1–20, simplify the expressions as much as possible, using rational exponent notation where appropriate.

1) $\sqrt{144}$

2) $\sqrt[3]{-64}$

3) $\sqrt{\dfrac{1}{9}}$

4) $\sqrt[5]{-32}$

5) $\sqrt{\dfrac{4}{49}}$

6) $\sqrt[3]{\dfrac{8}{27}}$

7) $8^{\frac{5}{3}}$

8) $(-8)^{\frac{5}{3}}$

9) $(-32)^{\frac{2}{5}}$

10) $-(32)^{\frac{2}{5}}$

11) $\left(\dfrac{16}{9}\right)^{-\frac{3}{2}}$

12) $(.01)^{-\frac{3}{2}}$

13) $(.008)^{\frac{4}{3}}$

14) $2^{\frac{5}{2}} 2^{\frac{3}{2}}$

15) $8^{\frac{5}{3}} 4^{\frac{1}{2}}$

16) $\left(3^{2/3}\right)^{3/4}$ 17) $\dfrac{2^{1/7}}{2^{3/2}}$ 18) $3\left(27^{2/3}\right)^{5/2}$ 19) $\dfrac{\left(2^{1/3}\right)^{2/5}}{\sqrt[5]{2}}$ 20) $3^{1/3}9^{1/3} + 2^{1/3}16^{1/6}$

21)   If $a > 0$ and $b > 0$, is $\sqrt{a^2 b^2}$ the same as $a\,b$? Justify your answer.

22)   If $a > 0$ and $b > 0$, is $\sqrt{a^2 + b^2}$ the same as $a + b$? Justify your answer.

23)   Is $\sqrt[3]{x^3 - 8}$ the same as $x - 2$? Justify your answer.

24)   If $x^2 + y^2 = 25$, can we conclude that $x + y = 5$? Why or why not?

25)   If $a > 0$ and $b > 0$, is $\sqrt{a + b}$ the same as $\sqrt{a} + \sqrt{b}$? If yes, prove it. If no, find a and b where two expressions are not the same.

1.6 Percent

Percent, represented by the symbol %, means "per hundred." So 5% of 400 means 5 one-hundredths of 400, which is $\left(\dfrac{5}{100}\right)(400) = 20$. In general, x % of y is $\left(\dfrac{x}{100}\right)(y)$.

Example 1:   Find 15% of 90.

Solution: $\dfrac{15}{100} \cdot 90 = \dfrac{15 \cdot 9}{10} = \dfrac{135}{10} = 13.5$ ∎

Example 2:   Find 1% of 320.

Solution: $\dfrac{1}{100} \cdot 320 = \dfrac{320}{100} = 3.2$

Note that 1% of any number is always $\dfrac{1}{100} = .01$ of the number, so move the decimal point over two places to the left! (For example, 1% of 4567 is 45.67 and 1% of 378.2 is 3.782 .) ∎

Example 3:   Find $\frac{1}{3}$% of 930.

Solution: Since 1% of 930 is 9.3, $\frac{1}{3}$% of 930 would be $\frac{1}{3}$ of 9.3 or 3.1.

Alternatively: $\dfrac{\frac{1}{3}}{100} \cdot 930 = \dfrac{930}{300} = \dfrac{93}{30} = \dfrac{31}{10} = 3.1$ ∎

Example 4:   Find 250% of 150.

Solution: $\dfrac{250}{100} \cdot 150 = \dfrac{250 \cdot 15}{10} = 25 \cdot 15 = 375$ ∎

Example 5:   A CD usually sells for $15.99. Shower Records is having a big sale offering 60% off. How many CD's can you get for $20, and exactly how much would you pay (assuming no sales tax)?

Solution: A 60%-off sale means that the CD's will cost 40% of their original price. Now 40% of $15.99 is

$$\dfrac{40}{100} \cdot (\$15.99) = \dfrac{4 \cdot (\$15.99)}{10} = \$6.396 = \$6.40 \quad \text{(rounded up to the}$$

nearest cent). So you can buy three CD's at a cost of $19.20. ∎

Example 6:   The new federal budget calls for a 30% cut in the deficit to $147 billion. What would the deficit have been if it had not been cut?

Solution: Let x = the uncut deficit.

So 70% of x is 147 billion – that is

$$\frac{70}{100} \cdot x = 147 \text{ billion.}$$

Solving for x: $x = \frac{100}{70} \cdot (147 \text{ billion})$

$$= 210 \text{ billion.}$$

So the deficit would have been 210 billion dollars. ■

Exercises 1.6

1) Find 25% of 200.   2)   Find 3.3% of 7.1.

3) The number 587 is 45% of what number?

4) The bookstore is having an inventory sale with a 25% reduction in prices. How much will a $16.99 sweatshirt cost you? How about a $1499.00 computer?

5) After a 40% cost reduction, a textbook that you purchased cost $20.40. What was the original price?

6) A hardware store is having a sale with a 20% reduction in prices. How much will a $239 table saw cost you?

7) After a 65% cost reduction, your new pet bird cost $84. What was the original price?

8) Joe heard that the government was selling foreclosed land at 35% off. He went to look at a farm going for the reduced price of $97,500. Joe bought the farm. How much did he save?

9) Diana borrowed $100,000 to start a new business. The loan has an interest rate of "prime plus two," with only the interest to be paid once a year. (Nice if you can get it.) The prime rate in the first year was 6%, and so her interest rate was 8%.

a) How much interest did she pay at the end of the first year?

b) The prime rate in the second year rose to 9%. How much did she pay after the second year, assuming the principal remains untouched?

c) How much did the interest payment increase from the first year to the second?

d) What was the percentage increase?

1.7 Scientific Notation, Calculators, Rounding

Decimal notation is OK for many numbers, but for really large or small numbers it is a big pain. Take the deficit (please!): 210 billion = 210,000,000,000 (in decimal form). But notice that all the zeros really represent multiplication by 10 so that

$$210,000,000,000 \;=\; 21 \times 10^{10} \;=\; 2.1 \times 10^{11}.$$

This last expression using the factors of 10 is called <u>scientific notation</u>. Notice that scientific notation calls for:

$$\pm \begin{bmatrix} \text{a number greater than or equal} \\ \text{to 1, but less than 10, written} \\ \text{in decimal form} \end{bmatrix} \times 10^{\left(\begin{array}{c} \text{some integer, positive} \\ \text{or negative, or 0} \end{array} \right)}$$

For example, you would write .0000536 in scientific notation as 5.36×10^{-5}. You would write 6437.8 as 6.4378×10^3. Just count the number of positions you have to move the decimal point so that the resulting number is between 1 and 10. If you have to move the decimal point left n places, the exponent is n; if you move it n places to the right, the exponent is $-n$.

<u>Remark</u>:   On most calculators, very large or small numbers are given in something like scientific notation. For example, the number 3.28×10^{-9} may look something like 3.2800000 E -09, or 3.28 $^{-09}$.

Example 1:   The number of molecules in a mole (a mole??) of gas is called Avogadro's number. It is equal to 6.023×10^{23} . Write it in decimal [nonscientific] form.

Solution: $6.023 \times 10^{23} \;=\; 602,300,000,000,000,000,000,000$

See why scientific notation is so handy!   ■

Example 2:   The speed of light (in a vacuum) is 186,000 miles per second. How big is a light-year (the distance light travels in 1 year)? Express the answer in scientific notation, rounded to three digits.

Solution: First, the speed of light written in scientific notation is 1.86×10^5. Now there are 365 days a year, so there are

365 days × 24 (hr/day) = 8760 hours per year, and

365 (days/year) × 24 (hr/day) × 3600 (sec/hr) = 31,536,000 seconds per year. In scientific notation, this number is 3.1536×10^7 seconds.

As you know, in each of these seconds, the light travels 186,000 miles. So in 1 year, it travels (31,536,000)(186,000) miles, and hence

$$
\begin{aligned}
\text{1 light-year} \quad &= \quad (\, 3.1536 \times 10^7 \,)(\, 1.86 \times 10^5 \,) \\
&= \quad (\, 3.1536 \,)(\, 1.86 \,) \times (\, 10^7 \,)(\, 10^5 \,) \\
&= \quad 5.865696 \times 10^{12} \\
&\cong \quad 5.87 \times 10^{12} \ \text{miles.} \ \blacksquare
\end{aligned}
$$

Remark 1:   This is in scientific notation since the first factor (5.87) is between 1 and 10. If it had not been between 1 and 10, you would have to write it differently. For example, if the product had ended up being 22.87×10^{12}, you would have to write it as 2.287×10^{13}.

Remark 2:   Notice that when the number was <u>rounded</u> to 3 digits, the result was 5.87, not 5.86. That's because the next digit was ≥ 5, and so 5.87 is a closer approximation to the actual value than 5.86. Right?

Scientific and graphing calculators have a key called y^x. If you're calculating interest, that key is a lifesaver.

Example 3:   a) Calculate $(1.07)^{10}$.

b) If you invest \$1000 at 7 % interest, compounded annually, what is the value of your investment in 10 years? (Round your answer to the nearest penny.)

Solution: a) Enter: 1.07

Press: y^x

Enter: 10

Press: =

The result on your calculator should be 1.967151357. It's probably unnecessary and a big nuisance to have all those digits. It may be fine to round it to 1.967 or 1.97, depending on what you are using it for.

b) Well, after 1 year the value would be

$$1000 + 7\% \text{ of } 1000 = 1000 + (0.07)(1000)$$
$$= 1000\,(1.07).$$

So after 2 years, it would be

$$[1000\,(1.07)]\,(1.07) = 1000\,(1.07)^2.$$

Similarly, after 10 years it would be

$$1000\,(1.07)^{10} = \$1967.15,$$

where the answer is rounded to the nearest penny. ∎

Exercises 1.7

1) Express the following numbers in scientific notation, rounded to three digits.

    a) 382935.9938   b) -0.000724 c) 3.000001   d) 200.001

2) Compute the following and express in scientific notation, rounded to three digits. You may use your calculator.

    a) $(2.35 \times 10^5) \times (4.032 \times 10^2)$ b) $(-6.15 \times 10^{-2}) \times (5.032 \times 10^6)$

    c) $(-5.001 \times 10^{-2}) \times (-7.001 \times 10^{-99})$ d) $\dfrac{3.24 \times 10^2}{4.23 \times 10^3}$

    e) $\dfrac{-1.33 \times 10^{-2}}{7.9 \times 10^5}$ f) $(3.82 \times 10^{-1})^3$

3) a) If you invest $ 2500, making 6 % annually for 20 years. What will its value be?

    b) If the cost of living increased at a constant rate for these 20 years at 4 % annually, then the "real" value of your investment, meaning your money's purchasing power, increases at 2 % annually. What is the increase in purchasing power in 20 years? What is the percentage increase?

4) a) If the $ 24 that bought Manhattan Island on May 6, 1626, had been invested at 5 %, what would it be worth on May 6, 1998? Express the answer in scientific notation, rounded to 2 digits.

    b) If the interest rate had been 6 %, what would be the value of the investment in 1998?

    c) Roughly speaking, the 1998 value at the 6 % rate was how many times as big as the 1998 value at the 5 % rate?

    d) What is the percentage increase?

1.8 Intervals

An <u>interval</u> is just a connected piece of the number line. For example, the set of all real numbers between 0 and 1, including 0 and 1 is called a <u>closed interval</u> and is denoted [0,1].   If you don't mean to include 0 and 1, you have an <u>open interval</u> and express it by using round parentheses, (0,1).   If 0 is to be included, but not 1, you write [0,1). Each of these intervals can also be shown on the number line, or expressed as a pair of inequalities. That is:

Interval Notation	Number Line	Inequalities

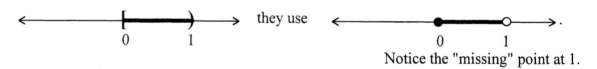

[0,1]		$0 \leq x \leq 1$
(0,1)		$0 < x < 1$
[0,1)		$0 \leq x < 1$
(0,1]		$0 < x \leq 1$

<u>Note:</u>   Some books use a different notation for an interval on the number line. Instead of

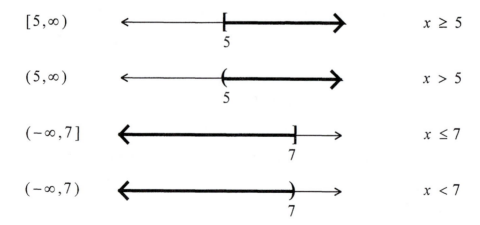

they use

Notice the "missing" point at 1.

All of the above are called <u>finite intervals</u> because their length is finite. There are also <u>infinite intervals</u>. Here are some examples.

$[5,\infty)$		$x \geq 5$
$(5,\infty)$		$x > 5$
$(-\infty,7]$		$x \leq 7$
$(-\infty,7)$		$x < 7$

Definition: Let A and B be two sets of objects of any sort.

a) The set of all objects that are in **both** A and B is called " A intersection B," and is denoted $A \cap B$.

b) The set of all objects that are in **either A or B or both** is called " A union B," and is denoted $A \cup B$.

The above definitions can be applied to intervals, and they often are. Consider the following:

Example 1: a) $(-\infty,5) \cap (3,\infty)$ is the set of all numbers that are less than 5 and at the same time greater than 3. So it is the set $(3,5)$.

b) Similarly, $(-\infty,10] \cap (-1,\infty) = (-1,10]$.

c) $(-\infty,10) \cap (21,\infty) = \varnothing$, the empty set, since there are no numbers that are both less than 10 and at the same time greater than 21.

d) $(-\infty,5) \cup (4,\infty) = (-\infty,\infty)$, the complete number line.

e) $(-\infty,5] \cup (10,\infty)$ can't be written in a more compact form. ∎

In the last example we see the use of union to describe sets of points on the number line which are not connected. It is often the case that we speak of a set of numbers consisting of two or more intervals, finite or infinite. This is when the union sign really comes in handy!

Exercises 1.8

1) Represent the following sets of numbers using interval and number line notation.

a) $-1 \le x \le 3$ b) $-1 < x \le 3$ c) $-3 \le x < 1$

d) $-3 \le x \le 4$ e) $-\dfrac{1}{2} < x \le \sqrt{2}$ f) $\pi \le x \le 5$

g) $0 < x$ h) $3 \le x$ i) $x < -4$

j) $3-\pi \le x$ k) $x < 5$ l) $x \le 3$

2) Represent the following intervals using inequalities.

   a) $(3,7)$ b) $(-4,-1]$ c) $(-\infty,19]$

   d) $[2,10)$ e) $[-2,-1]$

3) Simplify if possible:

   a) $(-\infty,5) \cap [3,\infty)$ b) $(-\infty,5) \cup [3,\infty)$

   c) $(-\infty,-2) \cap [-2,\infty)$ d) $(-\infty,\infty) \cap [4,7]$

4) Consider the two intervals $(-7, 1]$ and $(3,7)$. Represent each of these intervals on the number line. Represent the union of these two sets, that is $(-7,1] \cup (3,7)$, on the number line. Can you simplify the union using only one interval?

Chapter 2

Solving Equations

2.1 Equations of Degree 1 (Linear Equations)

You're asking: why are we doing this? This is the easiest thing in algebra. Well, it is, but it can get confusing when too many different variables are floating about. As long as you keep things straight, it really _is_ easy. We'll start off simple, and each new example will bring up some new twist and how to deal with it.

The expression $ax + b$, with $a \neq 0$, is a polynomial of degree 1, and so the equation $ax + b = 0$ is called an <u>equation of degree 1</u>. Since the graph of the function $ax + b$ is a straight line (See Section 3.3), the equation $ax + b = 0$ is also called a <u>linear equation</u>.

Example 1: Solve for x: $\qquad 4x - 16 = 0$.

Solution: First, get rid of the -16 by adding 16 to both sides:

$$4x - 16 + 16 = 0 + 16,$$

which gives $\qquad 4x = 16$.

Now, get rid of the 4 in front of the x by dividing both sides by 4:

$$\frac{4x}{4} = \frac{16}{4},$$

giving $\qquad x = 4$. Easy!

It's always a good idea to check your answer by substitution. That is, if we substitute $x = 4$ back into the original equation, it works! It is certainly true that

$$4 \cdot 4 - 16 = 0. \; \blacksquare$$

Example 2: Solve for x: $\dfrac{2}{3}x + 1 = 0$.

Solution: Get rid of the 1 by subtracting it from both sides. This gives

$$\frac{2}{3}x + 1 - 1 = 0 - 1$$

or $$\frac{2}{3}x = -1.$$

Now, to isolate the variable x we need to divide by $\dfrac{2}{3}$. Instead, just like in Chapter 1, we multiply by its reciprocal, $\dfrac{3}{2}$:

$$\frac{3}{2} \cdot \frac{2}{3}x = \frac{3}{2} \cdot (-1).$$

This gives $$x = -\frac{3}{2}. \quad \blacksquare$$

Example 3: Solve for y: $\sqrt{2}\, y - 4 + \sqrt{2} = 0$.

Solution: While this looks a little more complicated, y appears in only one place so you can start "peeling the onion." This is what we will call the method of stripping off terms, or factors, until the desired variable is exposed.

In this case, you first peel away the $-4 + \sqrt{2}$ by subtracting it from both sides to get

$$\sqrt{2}\, y = -(-4 + \sqrt{2})$$

or $$\sqrt{2}\, y = 4 - \sqrt{2}.$$

The next layer to peel away is that $\sqrt{2}$ on the left side of the equation. Divide by $\sqrt{2}$ to get

$$y = \frac{4 - \sqrt{2}}{\sqrt{2}}.$$

We can simplify this answer as follows:

$$y = \frac{4 - \sqrt{2}}{\sqrt{2}} = \frac{4}{\sqrt{2}} - \frac{\sqrt{2}}{\sqrt{2}} = 2\sqrt{2} - 1. \quad \blacksquare$$

In each of the preceding examples we obtained a solution by peeling the onion. The steps taken in stripping away the layers to expose the desired variable involved addition, subtraction, multiplication, or division of real numbers. Things can get a little more confusing when the equation involves more than the one variable you are solving for, but the idea is exactly the same. So, keep in mind what you have learned, and forge ahead!

Example 4:   Solve for x: $\qquad y^2 x + w^2 = 0.$

Solution: Now the equation has several variables. Remember the variable you wish to solve for and concentrate on isolating that one variable. Here, we wish to solve for x. Even though things seem a little more complicated, x still appears in only one place, so you can start peeling. In this case, first peel away the w^2 by subtracting it from both sides to get

$$y^2 x = 0 - w^2$$

or $\qquad y^2 x = -w^2.$

Next peel off the y^2 by division:

$$\frac{y^2 x}{y^2} = \frac{-w^2}{y^2},$$

resulting in $\qquad x = \dfrac{-w^2}{y^2}. \quad \blacksquare$

Again, we solve the equation by stripping away the layers to expose the desired variable. Unlike the first three examples, the last example also had variables, w, and y, in addition to the variable we were trying to solve for, x. These variables, however, just stand for real numbers, and so the mathematical steps taken to peel them away are exactly the same as those taken in the other examples.

Example 5:   Solve for x: $2y^4 x + (yw^2 - 2)^2 = 0.$

Solution: Again, here the x appears in only one place, so you can start peeling. In this case, first peel away the $(yw^2 - 2)^2$ by subtracting it from both sides to get

$$2y^4 x = 0 - (yw^2 - 2)^2$$

or $$2y^4 x = -(yw^2 - 2)^2.$$

Next peel off the $2y^4$ by division:

$$\frac{2y^4 x}{2y^4} = \frac{-(yw^2 - 2)^2}{2y^4},$$

which results in $$x = -\frac{(yw^2 - 2)^2}{2y^4}. \blacksquare$$

Sometimes the variable you wish to solve for occurs more than once. In many cases this is not a problem because you can rewrite the equation in such a way that the variable occurs only once, and then solve as usual.

Example 6:   Solve for x: $4x + 3 = 2x + 1.$

Solution: Here x occurs more than once, but if we subtract $2x$ from both sides, we get

$$2x + 3 = 1.$$

Notice that we have, in effect, moved all the terms containing x to the left side of the equation. Now, we subtract 3 from both sides to get

$$2x = -2,$$

where all the terms <u>not</u> containing x are now moved to the right side.

Dividing by 2 gives the final answer:

$$x = \frac{-2}{2} = -1. \blacksquare$$

Example 7:   Solve for x: $2x + 5y = 3x + y + 1$.

Solution: Even though this appears more complicated because of the extra y-terms floating around, it really isn't any more difficult. Here, the x occurs more than once, so we subtract $3x$ from both sides, to get

$$-x + 5y = y + 1.$$

Notice that the terms containing x are all on the left side.

Now, we subtract $5y$ from both sides:

$$-x = y + 1 - 5y,$$

which gives $-x = 1 - 4y$.

Notice that the terms not containing x are all on the right side.

Dividing by -1 gives the final answer:

$$x = \frac{(1 - 4y)}{-1}$$

$$= \frac{1}{-1} - \frac{4y}{-1}$$

$$= -1 - (-4y)$$

$$= -1 + 4y$$

$$= 4y - 1. \ \blacksquare$$

Example 8:   Solve for y: $\dfrac{2}{3}y + 2x - 1 = \dfrac{3}{4}y + x - \dfrac{1}{2}$.

Solution: Here y occurs more than once, but we can proceed as in the last example. First we subtract $2x - 1$ from both sides:

$$\frac{2}{3}y = \frac{3}{4}y + x - \frac{1}{2} - (2x - 1),$$

which is the same as

$$\frac{2}{3}y \;=\; \frac{3}{4}y + x - \frac{1}{2} - 2x + 1,$$

or
$$\frac{2}{3}y \;=\; \frac{3}{4}y + x - 2x + 1 - \frac{1}{2},$$

and finally
$$\frac{2}{3}y \;=\; \frac{3}{4}y - x + \frac{1}{2}.$$

Now, we can combine the two terms that contain y by subtracting $\frac{3}{4}y$ from both sides to get

$$\frac{2}{3}y - \frac{3}{4}y \;=\; \frac{3}{4}y - x + \frac{1}{2} - \frac{3}{4}y.$$

Canceling the $\frac{3}{4}y$'s on the RHS (right-hand side), and combining the y-terms on the LHS gives:

$$\left(\frac{2}{3} - \frac{3}{4}\right)y \;=\; -x + \frac{1}{2}.$$

Remember how to subtract fractions: we get a common denominator, etc.

So
$$\frac{2}{3} - \frac{3}{4} \;=\; \frac{2 \cdot 4}{12} - \frac{3 \cdot 3}{12} \;=\; \frac{8}{12} - \frac{9}{12} \;=\; -\frac{1}{12}$$

and hence
$$-\frac{1}{12}y \;=\; -x + \frac{1}{2}.$$

Now, to expose the variable y, we divide by $-\frac{1}{12}$, that is, we multiply both sides by -12 to give the solution:

$$(-12)\cdot\left(-\frac{1}{12}\right)y \;=\; (-12)\cdot\left(-x + \frac{1}{2}\right),$$

or
$$y \;=\; (-12)\cdot(-x) + (-12)\cdot\left(\frac{1}{2}\right)$$

$$=\; 12x - 6. \;\blacksquare$$

Exercises 2.1

1)   Solve for x: $3x - 4 = 0$.

2)   Solve for x: $\dfrac{5}{12}x - 35 = 0$.

3)   Solve for z: $\dfrac{4}{3}z - 1 = \dfrac{1}{10}$.

4)   Solve for x: $\dfrac{2}{3}x - 3 = \dfrac{1}{5}x + 2$.

5)   Solve for y: $\dfrac{1}{2}y - \dfrac{1}{3} = \dfrac{1}{6} - 2y$.

6)   Solve for x: $3y^2 x + z^2 = 0$.

7)   Solve for y: $2zy + 3z + 1 = 0$.

8)   Solve for z: $2zy + 3z + 1 = 0$.

9)   Solve for x: $2y^2 x - y^2 - (1 + 3y) = x$.

10)   Solve for x: $x - y^2 + (2zy^2 + 2y)x = z^2 x + z$.

2.2 Equations of Degree 2 (Quadratic Equations)

The expression $ax^2 + bx + c$ with $a \neq 0$, is a polynomial of degree 2, and the equation $ax^2 + bx + c = 0$ is called an <u>equation of degree 2</u>. It is also called a <u>quadratic equation</u> because "quadratum" is Latin for "square." Notice that x occurs in two places, so you can't use the method of "peeling the onion." However, by using the method of completing the square, you can change the form of the equation in such a way that x occurs in only one place, and then peel away as usual. If you do that (see Appendix A), you will find that

$$x = \frac{-b \pm \sqrt{b^2 - 4ac}}{2a}.$$

This is the quadratic formula. It gives you the solution of that equation by simply substituting in the values of a, b, and c. No fuss, no bother, no completing the squares, no messy onion peels. It is one of the most useful formulas you will meet. **MEMORIZE IT NOW!!** If you don't, you will stray from the straight and narrow, and not even Elvis will be able to save you.

<u>Remark:</u>   The "$\pm$" sign in the quadratic formula means that there are possibly two solutions, one if we pick the "+" sign and the other if we pick the "−" sign. Sometimes the two solutions are, in fact, equal. (When does this happen?)

Example 1:   Solve for x: $x^2 - 3x + 2 = 0$.

Solution: Here $a = 1$, $b = -3$, and $c = 2$. So, according to the quadratic formula

$$x = \frac{-(-3) \pm \sqrt{(-3)^2 - 4 \cdot 1 \cdot 2}}{2 \cdot 1},$$

which simplifies to

$$x = \frac{3 \pm \sqrt{9 - 8}}{2} = \frac{3 \pm 1}{2}.$$

When the "+" sign is used, we get the solution

$$x = \frac{3 + 1}{2} = \frac{4}{2} = 2.$$

When the "−" sign is used, we get the solution

$$x = \frac{3-1}{2} = \frac{2}{2} = 1.$$

So there are two solutions: $x = 1$ and $x = 2$. Can you see how useful this formula is? ∎

Example 2:   Solve for x: $2x^2 - 3x + 1 = 0$.

Solution: Here $a = 2$, $b = -3$, $c = 1$, and $b^2 - 4ac = 1$. So, according to the quadratic formula

$$x = \frac{-(-3) \pm \sqrt{(-3)^2 - 4 \cdot 2 \cdot 1}}{2 \cdot 2},$$

which simplifies to

$$x = \frac{3 \pm \sqrt{9 - 8}}{4} = \frac{3 \pm 1}{4}.$$

The "+" sign gives $x = \frac{3+1}{4} = \frac{4}{4} = 1.$

The "−" sign gives $x = \frac{3-1}{4} = \frac{2}{4} = \frac{1}{2}.$

So there are two solutions: $x = 1$ and $x = \frac{1}{2}$. ∎

Example 3:   Solve for s: $s^2 + 4s + 4 = 0$.

Solution: Don't let the s fool you, this is just a quadratic equation in s, and s takes the place of x in the preceding discussion. Here $a = 1$, $b = 4$, and $c = 4$. So,

$$s = \frac{-(4) \pm \sqrt{(4)^2 - 4 \cdot 1 \cdot 4}}{2 \cdot 2}$$

$$s = \frac{-4 \pm \sqrt{16 - 16}}{2} = \frac{-4 \pm 0}{2} = -2.$$

In this case there is only one root. When this happens, the root is called repeated, or a root of order 2. (When speaking of equations, root is another word for solution.) It means that the original quadratic could be factored as $s^2 + 4s + 4 = (s + 2)^2$, and yes, you guessed it: Repeated roots occur when the quadratic is a perfect square. ∎

Sometimes you can solve a quadratic equation by "peeling the onion" more quickly than using the quadratic formula. This happens when the coefficient $b = 0$. Consider the next example.

Example 4:   Solve for x: $x^2 - 9 = 0$.

Solution: Here $a = 1$, $b = 0$, and $c = -9$. Instead of using the quadratic formula we will "peel the onion." First we add 9 to both sides of the equation to get

$$x^2 = 9.$$

Now we take square roots of both sides of the equation, remembering that all positive numbers have two square roots. So

$$\pm x = \pm 3,$$

which means $x = 3$ or $x = -3$. ∎

Sometimes a quadratic equation can't be solved in terms of real numbers. Consider the following.

Example 5:   Solve for y: $y^2 + 2y + 2 = 0$.

Solution: Here $a = 1$, $b = 2$, and $c = 2$, so that

$$y = \frac{-2 \pm \sqrt{4-8}}{2} = \frac{-2 \pm \sqrt{-4}}{2} = \frac{-2 \pm 2\sqrt{-1}}{2} = -1 \pm \sqrt{-1}.$$

The term $\sqrt{-1}$ is not defined as a real number because no real number when squared gives a negative number. So there are no real solutions. ■

If $b^2 - 4ac < 0$, as in the last example, there are no real solutions, meaning that there are no solutions that are real numbers.. But wishful thinking is a great motivator, and in this spirit, mathematicians several centuries ago agreed to consider $\sqrt{-1}$ an "imaginary number." We represent $\sqrt{-1}$ by the symbol i (for imaginary). In this case the roots of the last equation could be written as $-1 \pm i$. Any number of the form $a + ib$ is called a <u>complex number</u>, and all four basic operations (addition, subtraction, multiplication, and division) carry over from real numbers, provided we remember $i^2 = -1$. (In the 1800s, complex numbers were removed from the realm of hocus-pocus and put onto a logically sound foundation – which means that today we can all sleep much easier.) Now check the numbers $-1 \pm i$ to see whether they are in fact solutions to the problem in Example 5.

Example 6: Verify that $y = -1 \pm i$ are solutions to $y^2 + 2y + 2 = 0$.

Solution: We will check each value, one at a time.

a) If $y = -1 + i$, then

$$y^2 = (-1+i)^2 = (-1)^2 - 2i + i^2$$
$$= 1 - 2i - 1$$
$$= -2i$$

$$2y = 2(-1+i) = -2 + 2i.$$

Hence $y^2 + 2y + 2 = -2i - 2 + 2i + 2 = 0$. So it works!

b) If $y = -1 - i$, then

$$y^2 = (-1-i)^2 = (-1)^2 + 2i + (-i)^2$$
$$= 1 + 2i - 1$$
$$= +2i$$

$$2y = 2(-1-i) = -2 - 2i.$$

Hence $y^2 + 2y + 2 = 2i - 2 - 2i + 2 = 0$. ■

<u>Remark:</u>   These examples demonstrate the three possible outcomes when solving quadratic equations. The number $b^2 - 4ac$ is called the <u>discriminant,</u> since it allows you to "discriminate" among the following three cases that indicate what types of roots you will have.

1) $b^2 - 4ac > 0$:   In this case, the quadratic equation has two real and distinct roots – that is, two roots that are different numbers. This was the situation in Example 1.

2) $b^2 - 4ac = 0$:   Here, the square root in the solution disappears, leaving one real repeated root. This happened in Example 3.

3) $b^2 - 4ac < 0$:   In this last case, the roots are complex numbers, as was the case in Example 5.

It is possible that in your travels you will meet up with a quadratic equation in x where the coefficients a, b, and c are functions of some other variables. Take the following example.

Example 7:   Solve for x: $zx + 2yx^2 + yx + z^2 - y^2 = 0.$

Solution: On first appearance this seems formidable, but by writing the x -term first, then the x-term, and then the term without x, we get

$$(2y)x^2 + (z+y)x + (z^2 - y^2) = 0.$$

We can that see this is a quadratic equation in x with $a = 2y$, $b = z+y$, and $c = z^2 - y^2$. Hence substituting these values into the quadratic formula gives

$$x = \frac{-(z+y) \pm \sqrt{(z+y)^2 - 8y(z^2 - y^2)}}{4y}. \blacksquare$$

Exercises 2.2   Find the solutions, both real and complex, for Exercises 1–18.

1) $x^2 + 5x + 4 = 0$

2) $g^2 - 2g - 8 = 0$

3) $x^2 + 6x + 9 = 0$

4) $2y^2 + y - 1 = 0$

5) $y^2 - 8y + 16 = 0$

6) $\dfrac{x^2}{4} + 2x + 1 = 0$

7) $y^2 - 2y + 2 = 0$

8) $3y^2 - 4 = 0$

9) $x^2 + 3x = 4$

10) $x^2 + 7x + 4 = 0$

11) $g^2 - 4g - 5 = 0$

12) $8x^2 + 6x + 1 = 0$

13) $2y^2 + 7y + 4 = 0$

14) $y^2 - \dfrac{2y}{15} - \dfrac{1}{15} = 0$

15) $\dfrac{x^2}{3} + 2x - 1 = 0$

16) $y^2 + 3 = 0$

17) $x^2 + 2x = 4$

18) $3y^2 - 1.8y - 1.2 = 0$

19)   Solve $\dfrac{1}{2}x^2 + yx - y^2 = 0$ for x.

20)   Solve $\dfrac{1}{4}x^2 - (y + z)x + z^2 + y^2 = 0$ for x.

21)   Solve $\dfrac{1}{4}x^2 - (y + z)x + z^2 + y^2 = 0$ for z.

2.3 <u>Solving Other Types of Equations</u>

Some of the methods that we have used to solve equations of degree 1 and 2 can also be used to solve other kinds of equations. There are four basic tools in our equation solving toolbox:

1) <u>Peeling the Onion</u>; for any equation where the variable that we wish to solve for occurs in only one place, or any equation that can be put into that form.

2) <u>The Quadratic Formula</u>; for equations that are of "quadratic type", meaning that a well-chosen substitution can change the given equation into an ordinary quadratic equation.

3) <u>The Zero-Factor Property</u>; which allows us to solve equations of the form $f(x) = 0$, by factoring $f(x)$, and setting each factor equal to zero.

4) <u>Graphing Calculator Techniques</u>; which can be used to solve the equation $f(x) = 0$, by graphing $f(x)$, and seeing where the graph crosses the x-axis. Those places are the solutions of the equation. (Consult the manual that came with your calculator to see how to graph functions effectively.)

Before we start, we must emphasize the need to check the calculated solutions to make sure that they truly are solutions by substituting them into the original equation. Sometimes when manipulating an equation to find a solution we run the risk that the "solutions" we calculated do not satisfy the original equation. Such "solutions" are called <u>extraneous</u>. Here are some examples of when extraneous solutions can creep into a problem.

a) Whenever we square both sides of an equation. For example, if we square both sides of the equation $x = 2$, we get $x^2 = 4$. The second equation has a solution $x = -2$, which clearly is not a solution to the original equation.

b) When we have denominators, which of course can't be equal to zero. Consider $\dfrac{x-1}{x} = \dfrac{-1}{x}$. Multiplying both sides of this equation by x gives the equation $x - 1 = -1$, which has the solution $x = 0$. However, this solution does not satisfy the original equation.

c) If the given equation has terms in it like $\sqrt{x+1}$, $\log\left(x^2 - 1\right)$, or $(x-2)^{3/2}$, we must make sure that our "solution" is a number for which such terms are defined.

d) In word problems, we must make sure that calculated solutions are physically reasonable. For example, if a word problem leads to the result that the length of a rectangle is -3 or $+5$, we must ignore the solution -3, since it does not make sense to have a negative length.

Let's begin by examining some problems of the onion peeling variety.

Example 1:   Solve for x: $\sqrt{x} - 3 = 0$.

Solution: This is not of degree 1, but it is of "onion type" because x is in one place only. To get at it, we remove the outermost layer by adding 3 to both sides to get

$$\sqrt{x} = 3.$$

Next, we remove the square root by squaring both sides to get

$$\left(\sqrt{x}\right)^2 = 3^2$$

or $$x = 9.$$

(Don't forget to check your solution!) ■

Example 2:   Solve for x: $(\sqrt{x}+2)^3 - 64 = 0$.

Solution: As in the last example, this equations is not degree 1 but it is of "onion type" because x is in one place only. To get at it, remove the outermost layer by adding 64 to both sides:

$$(\sqrt{x}+2)^3 = 64.$$

Next, remove the cube by taking the cube root of both sides of the equation,

$$\left((\sqrt{x}+2)^3\right)^{1/3} = 64^{1/3}$$

or $$\sqrt{x}+2 = 4.$$

Subtract 2 from both sides to give

$$\sqrt{x} = 2,$$

and then square both sides to get

$$x = 4.$$

Again, remember to check your solution! ∎

Example 3:   Solve $\dfrac{1}{x-5} + \dfrac{1}{x+5} = \dfrac{10}{x^2-25}$ for x.

Solution: This is not of onion type, but we can change that by multiplying by $x^2 - 25$ to get rid of the denominators. Doing so we get

$$\frac{x^2-25}{x-5} + \frac{x^2-25}{x+5} = 10.$$

By factoring and canceling we get

$$x + 5 + x - 5 = 10,$$

or $2x = 10$ and $x = 5.$

But $x = 5$ cannot be used in the original equation without dividing by zero. Hence there is no solution to this problem. Good thing we checked. ∎

Sometimes an equation can be reduced to a quadratic equation with a simple substitution. If an equation in x has three terms, one constant, one like x^m, and the other like x^{2m}, then the substitution $y = x^m$ leads to a quadratic equation in y. Consider the following example.

Example 4:   Solve for x: $x^4 - 5x^2 + 4 = 0.$

Solution: This is of quadratic type because if we use the substitution $y = x^2$, then this equation becomes

$$y^2 - 5y + 4 = 0,$$

which is quadratic and has roots $y = 1$ and $y = 4$. (Check this!) But 1 and 4 are NOT the solutions of the given equations.   Instead, we have $x^2 = 1$ or $x^2 = 4$, giving us the solutions $x = \pm 1$ and $x = \pm 2$. ∎

We have to be very careful, however, since taking square roots may give us solutions which are not real-valued. Consider the following example.

Example 5:   Solve for x: $x^4 - 5x^2 - 36 = 0$.

Solution: Again, letting $y = x^2$, we will obtain the quadratic equation

$$y^2 - 5y - 36 = 0,$$

which has roots $y = -4$ and $y = 9$. Since $y = x^2$, the solutions of the original equation are $x^2 = -4$ and $x^2 = 9$, giving $x = \pm 2i$ and $x = \pm 3$.

However, if only real solutions are asked for, the solutions are just 3 and −3. ∎

Example 6:   Solve $x^6 + 6x^3 - 16 = 0$ for real x.

Solution: Now we use the substitution $y = x^3$ to obtain

$$y^2 + 6y - 16 = 0,$$

which has solutions

$$y = -8 \quad \text{and} \quad y = 2.$$

So $x^3 = -8$ and $x^3 = 2$,

or $x = -2$ and $x = \sqrt[3]{2} = 2^{\frac{1}{3}}$. ∎

Example 7:   Solve $x + \sqrt{x} - 6 = 0$ for real x.

Solution: This is of quadratic type, since if we let $w = \sqrt{x}$ we get $w^2 = x$, and so

$$w^2 + w - 6 = 0.$$

The solutions of this quadratic are $w = 2$ or -3. (Do you agree?) Therefore $\sqrt{x} = 2$ or -3, where the -3 is obviously an extraneous solution. Hence $\sqrt{x} = 2$ or $x = 4$. Check it! ∎

For more complicated equations with coefficients that are also variables, the method is the same, but you should always verify (that is, check) the solutions.

The Zero-Factor Property (ZFP):

If a and b are real numbers, and $a \cdot b = 0$, then either $a = 0$, or $b = 0$, or both. (This result is used to solve equations where there are products of factors.) Consider the following.

Example 8:   Solve for x: $(x - 1)(x + 2) = 0$.

Solution: Here we can use the Zero-Factor Property. By associating $x - 1$ with a and $x + 2$ with b we see that this equation is equivalent to $a \cdot b = 0$. Applying the ZFP we deduce that either $a = 0$ or $b = 0$ (or both), and so

$$x - 1 = 0$$

or $$x + 2 = 0.$$

The solution to the first equation is clearly $x = 1$, and the solution to the second equation is $x = -2$. Hence the solutions to the original equation are $x = 1$ or $x = -2$. (Check this!) ∎

Notice that if we multiply the factors in the original equation in this example we get the quadratic equation $x^2 + x - 2 = 0$. Using the quadratic formula from the last section, we of course obtain the same solutions $x = 1$ or $x = -2$.

Example 9:   Solve for x: $(x^2 - 4)(x^2 + 2x - 3) = 0$.

Solution: Applying the ZFP gives the two equations

$$x^2 - 4 = 0$$

or $$x^2 + 2x - 3 = 0.$$

Each of these equations is quadratic and can be solved using the quadratic formula. The solutions to the first equation are $x = 2$ or $x = -2$. (Make sure you verify this for yourself!)

The solutions to the second equation are $x = -3$ or $x = 1$. (Check this also!) So the solutions to the original equation are: $x = 2$, $x = -2$, $x = -3$, or $x = 1$. ∎

The ZFP can be extended to cases when there are more than 2 factors. For example, if $a \cdot b \cdot c = 0$, then either $a = 0$, $b = 0$, or $c = 0$. When you have a polynomial equation $f(x) = 0$, the idea is to factor $f(x)$ as far as possible, and to then apply the ZFP. Remember that the right-hand side <u>must be zero</u> for the ZFP to apply. If your factoring skills are a little rusty, you can buff them up in Chapter 4.

Example 10:   Solve for x: $x^5 + 3x^4 - 4x^3 = 0$.

Solution: We can factor this by first taking out a common factor, and get

$$x^3 \left(x^2 + 3x - 4 \right) = 0.$$

Next, the second factor can be factored further to give

$$x^3 (x + 4)(x - 1) = 0.$$

The ZFP says that:

$$x^3 = 0,$$

$$x + 4 = 0,$$

or $$x - 1 = 0.$$

Hence $x = 0, -4$, or 1. Remember to check the solutions. Again, factoring methods are given in Chapter 4. ∎

Exercises 2.3   Find the solutions, both real and complex, for Exercises 1–16.

1) $(x-3)(x+6) = 0$ 2) $(x+2)(x-9) = 0$ 3) $(2x+1)(x-2) = 0$

4) $(x^2-9)(x+6) = 0$ 5) $(x^2-5)(x^2-16) = 0$ 6) $(x^2+1)\left(\dfrac{x}{3}-4\right) = 0$

7) $(x^2+4x-5)(3x^2-81) = 0$ 8) $z^4 - 8z^2 + 16 = 0$

9) $x^4 - 64 = 0$ 10) $s^4 - 9s^2 + 20 = 0$

11) $\sqrt{x} - 4 = 0$ 12) $\sqrt{x} - 3 = 5 - \sqrt{x}$

13) $(\sqrt{2x}-3)^3 + 1 = 0$ 14) $\left(\dfrac{1}{\sqrt{x}} - 3\right)^4 - 1 = 0$

15) $\dfrac{1}{\sqrt[3]{x^4}} = \dfrac{1}{\sqrt[3]{x}}$ 16) $\dfrac{4}{\sqrt{x^3}} = \dfrac{1}{\sqrt{x}}$

17)   Find all real solutions to $x^{2/5} - 3x^{1/5} + 2 = 0$.

18)   Find all solutions to $\dfrac{1}{x+1} + \dfrac{1}{x} = \dfrac{3}{x^2+x}$.   (Hint: $x^2 + x$ is a common denominator for the two algebraic fractions on the LHS.)

19)   Find all solutions to $\dfrac{1}{x-1} + \dfrac{1}{x+1} = \dfrac{2}{x^2-1}$.

20)   Find all real solutions to $\dfrac{1}{x-4} + \dfrac{1}{x+4} = \dfrac{4}{x^2-16}$.

21)   Find all real solutions to $x^6 - 26x^3 - 27 = 0$.

Chapter 3

Introducing Functions and Graphs

3.1 Functions and Their Representations

In calculus, we are concerned with quantities whose value is not constant, but rather depends on the value of some other quantity, often time. For example, the temperature displayed in Times Square in New York City changes quite a bit, but it has a particular value at any given point in time. We say the temperature is a <u>function</u> of time.

Example 1: Here is the recording of the temperature in Times Square, N.Y.C. between 8 AM and 8 PM on a certain day:

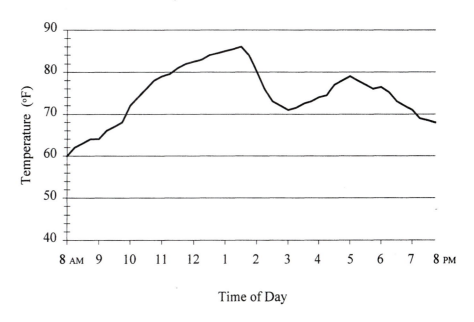

Time of Day

a) Approximately what was the temperature at noon? b) What was the highest temperature, and when did it occur? c) What was the lowest, and when did it occur? d) Give a rough estimate of the average temperature over that 12-hour interval. e) What possible explanations could there be for the dip in the temperature in midafternoon.

Solution: a) At noon, the temperature was roughly 83° F.

b) The highest temperature was about 86° F at 1:30 PM.

c) The lowest temperature was about 60° F, at 8 AM.

d) The average was obviously less than 80° F and greater than 70° F, maybe around 75° F.

e) Here are some possibilities: a passing shower, shade from a tall building, King Kong blocking the sunlight. ■

The graph shown in the previous example is sometimes called a <u>line graph</u>, even though it is obviously not a line. Quantities that vary as a function of another quantity can always be represented by such a graph. Another way that a function can be represented is in the form of a table, and in using such a representation it is important to know how to translate the data into a graph.

Example 2:   The table below shows the price of a share of stock of B & M Inc. at the close of the trading day for the week of June 9 – 13, 1997. Take the function as given and display it by means of a graph.

Day	Closing Price of B&M Stock
Monday	$ 6.75
Tuesday	$ 7.75
Wednesday	$ 8.50
Thursday	$ 9.75
Friday	$ 11.25

Solution: Here, we use a <u>bar graph</u> to represent the stock price data.

Daily Closing Stock Price for B & M Inc.

Week of June 9 - June 13, 1997

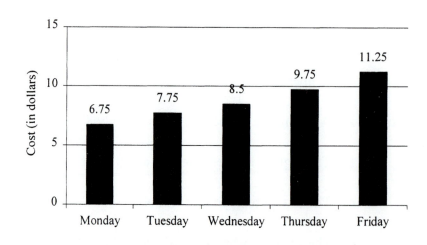

■

Sometimes, a function in tabular form can be represented as a line graph, as in the next example.

Example 3:   We're off to the post office to mail a letter. We check the cost of mailing a letter first-class and find the following table:

Weight not over (oz.)	Cost
1	$ 0.32
2	0.55
3	0.78
4	1.01
5	1.24
6	1.47

Here, the cost of mailing a first-class letter is a function of how much the letter weighs. Take the tabular representation of the function and graph cost as a function of weight.

Solution:

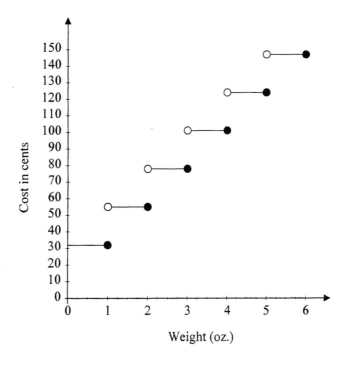

Notice that the open circles indicate missing points, while the solid circles indicate included points. For example, if the weight is exactly 1 ounce, the cost is 32 cents, not 55 cents. Also notice that this is a line graph, although a strange one. The function is said to be <u>piecewise defined.</u> ■

So we see that functions can be defined by tables and by graphs. They can also be defined by words and by symbolic expressions. This fact is sometimes referred to as the "Rule of Four." It is important that you can easily change functional representations from any one form to another. Consider the next example.

Example 4:   Hurts Car Rental charges $50 for a weekend, plus 15 cents per mile driven. Let C be the rental cost, in dollars, and let d be the distance driven, in miles. a) Express C symbolically as a function of d.   b) Graph C as a function of d.

Solution: a) $C = 50 +$ (number of miles driven)(0.15)

$C = 50 + (0.15)(d)$

b) To find the graph of C as a function of d, it is useful to make a table of values of d and the corresponding values of C, like the following.

d (in miles)	C (in dollars)
0	50
50	57.50
100	65
200	80
300	95
500	125
1000	200

Each line of the table yields a point in the graph. For example, the first line goes to the point $(0,50)$ in the d-C plane. Placing the point into the plane is called plotting the point. By plotting all of the points in this table we obtain the graph on the following page.

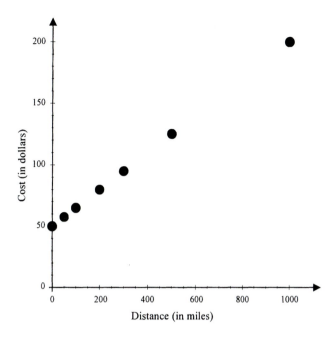

This gives you an incomplete graph, of course, because what if you drive 700 miles or 2000 miles? Well, you can plot those points too. Obviously you'll never finish the complete graph by only plotting points. Can you just draw a smooth curve through the points to get the complete graph? In this case, yes, but in some other cases, no. One of the strengths of Calculus is that it tells you whether drawing such a curve to get the complete graph is justified in any given case. In any event, using Calculus we can show that in this case it is justified, and so we get the following.

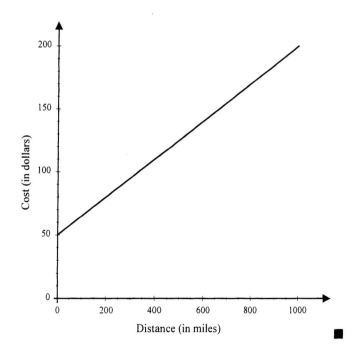

Exercises 3.1

1)   The volume, or total number of shares, traded on the New York Stock Exchange is given in the graph below for a six-week period during April and May of 1997.

N.Y.S.E. VOLUME (in millions of shares)

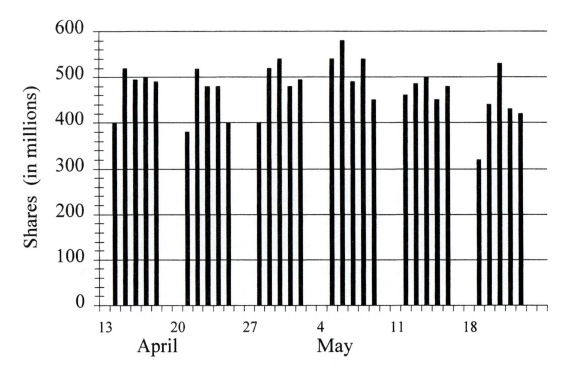

a)   Approximately how many millions of shares were traded on Monday, May 5th?

b)   What was the largest volume and when did it occur?

c)   What was the smallest volume and when did it occur?

d)   What was the average volume for the week of April 28th?

e)   Convert this graph to a table for the week of April 21st.

2) The population of sheep in Tasmania during the period spanning 1804–1924 is given in the following graph.

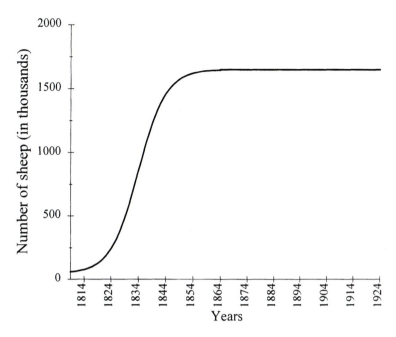

a) Approximately how many sheep were around in 1844?

b) Consider the statement: "From 1864 on, the curve is a horizontal line." Translate it into a statement about numbers of sheep.

c) Convert this graph to a table for the years 1814–1864 at 10-year intervals.

3) Your local Pizza Parlor sells a large pizza for $6 plus another $1.50 for each topping requested. The cost of a pizza as a function of how many toppings you get is given in the following table.

Number of Toppings	Cost of a Large Pizza
0	$ 6.00
1	$ 7.50
2	$ 9.00
3	$ 10.50
4	$ 12.00
5	$ 13.50

a) Represent pizza cost as a function of number of toppings with a graph.

b) Let C be the cost of a pizza, and let n be the number of toppings. Write down a symbolic expression that relates C and n.

3.2 Definitions and Notation

Let's get to the heart of the idea of functions. We have seen several different ways of presenting functions, using tables, words, algebraic expressions, and graphs. What they all have in common is given in the following definition.

Definition: a) Let g and x be two variable quantities which are related in such a way that to every value of x in a certain set A, there corresponds exactly one value of g, denoted $g(x)$. Then g is called a <u>function</u> of x, and A is called the <u>domain</u> of the function g.

                        b) If the domain is not specified, then it is assumed to be maximal. This means that the domain includes all x for which $g(x)$ makes sense.

                        c) As x takes on all the values in A, the set of all corresponding function values $g(x)$ is called the <u>range</u> of the function g.

Example 1:   Let g be the function defined by the equation $g(x) = \sqrt{x-2}$. Find its domain.

Solution: Since the domain is not specified, we know it is maximal, that is, it includes all values of x which we are allowed to use in the expression $\sqrt{x-2}$. Well, $\sqrt{x-2}$ has real values as long as $x - 2 \geq 0$. But this is equivalent to $x \geq 2$. Hence the domain is $[2, \infty)$. ∎

Example 2:   Let $k(x) = x^2 - 4$. Find its domain and range.

Solution: The domain includes all values of x for which $x^2 - 4$ is defined. Well, no matter what x is, we can always evaluate $x^2 - 4$. Hence the domain is all real numbers x, or $(-\infty, \infty)$. The range is the set of all possible values of $x^2 - 4$. Now think: since x^2 can be any number ≥ 0, $x^2 - 4$ can be any number we have ≥ -4. Therefore, the range is the set $[-4, \infty)$.

Note: The function $k(x) = x^2 - 4$ is a polynomial, and all polynomials have a domain of $(-\infty, \infty)$ because they can be evaluated at all real numbers. ∎

The preceding two examples had functions defined by symbolic expressions. Finding the domain and range of such functions can be a little tricky. When the function is defined by a table, it is much easier to determine the domain and range.

Example 3:   You're off to the local clothing store to buy some socks. You walk in the front door and right there, at the entrance, is a huge display table with the socks you love, and they're on sale. In the middle of the table is a sign:

Number of Pairs	Cost per Pair
1 – 5	$ 2.29
6 – 10	$ 1.99
11 or more	$ 1.79

Here, the cost of a pair of socks is a function of how many pairs you buy. What is the domain of this function, and what is the range? Also, how much would you spend (before tax) if you bought 9 pairs of socks?

Solution: To determine the domain, consider the number of pairs of socks a person could buy (assuming that they are going to buy socks). Since the number of pairs of socks is 1, 2, 3, 4, etc., the domain of this function is the set of all positive integers. Once you pick a value in the domain, the sign gives you the cost of the socks, that is, the function values. These are 1.79, 1.99, and 2.49, hence the range is the set {1.79, 1.99, and 2.49}.

If you buy 9 pairs of socks, then the cost per pair is $ 1.99, hence the total cost before tax is $ 17.91 . ■

Example 4:   Let $p(x) = x^2 + \dfrac{1}{2x - 1} - 1$.

Calculate a) $p(1)$, b) $p(2)$, c) $p(\frac{1}{2})$, d) $p(0)$, e) $p(x + h)$, f) $p(x + \Delta x)$, and g) $p(x + \pi - 17)$. (Note: Δx is a variable that just happens to looks a little different.)

Solution: a) $p(1) = 1 + \dfrac{1}{1} - 1 = 1$

b) $p(2) = 4 + \dfrac{1}{3} - 1 = \dfrac{10}{3}$

c) $p\left(\frac{1}{2}\right)$ does not exist, since the denominator would be 0. So we know that ½ is not in the domain of p.

d) $p(0) = 0 + \dfrac{1}{-1} - 1 = -2$

e) $p(x + h) = (x + h)^2 + \dfrac{1}{2(x + h) - 1} - 1$

$= x^2 + 2xh + h^2 + \dfrac{1}{2x + 2h - 1} - 1$

f) $p(x + \Delta x) = (x + \Delta x)^2 + \dfrac{1}{2(x + \Delta x) - 1} - 1$

$= x^2 + 2x\Delta x + (\Delta x)^2 + \dfrac{1}{2x + 2\Delta x - 1} - 1$

g) $p(x + \pi - 17) = (x + \pi - 17)^2 + \dfrac{1}{2(x + \pi - 17) - 1} - 1$ ∎

The last function we would like to talk about is the <u>absolute value function</u>. This function, denoted $|x|$, is defined piecewise as

$$|x| = \begin{cases} x & \text{if } x \geq 0 \\ -x & \text{if } x < 0 \ . \end{cases}$$

<u>Remark 1:</u>   Students sometimes think that $|x|$ is x "taken positively," whatever that means. Do not use that fuzzy notion, but rather use the actual definition shown above. Also notice that $-x$ may LOOK negative, it is NOT negative if $x < 0$. For example, if $x = -5$, then $-x = -(-5) = 5$. Right?

<u>Remark 2:</u>   Also notice that $\sqrt{x^2} = |x|$. (Verify this for $x = 3$ and $x = -3$.)

To graph the absolute value function we handle each piece from the definition, one line at a time. If $x \geq 0$, then $|x| = x$. We know the graph of $y = x$ is just the 45° line through the origin, and so that's the graph of $|x|$ for $x \geq 0$. See Figure 1(a) below.

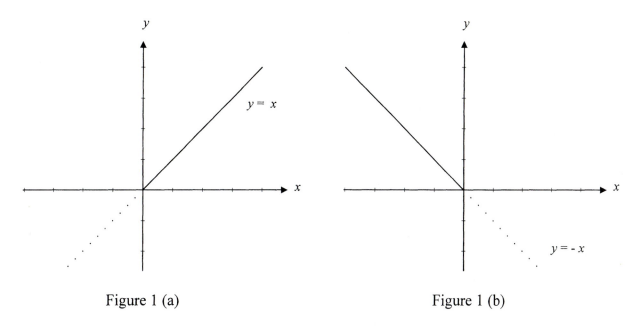

Figure 1 (a)   Figure 1 (b)

If $x < 0$, then $|x| = -x$, whose graph is shown in Figure 1 (b) above.

Now let's put these two graphs together; the solid piece from Figure 1(a) for $x \geq 0$ and the solid piece from Figure 1(b) for $x < 0$:

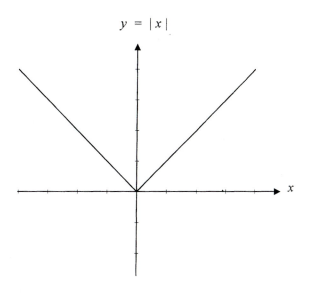

Figure 2

Projecting the graph onto the x-axis shows that the domain is $(-\infty, \infty)$. Projecting the graph onto the y-axis shows that the range is $[0, \infty)$.

Exercises 3.2

1) Suppose you're shopping for compact discs. A local store has a sale on used CD's. The cost is written up as:

Number of CD's	Cost per CD
1 – 3	$ 7.99
4 – 7	$ 6.99
8 – 10	$ 6.49
11 or more	$ 5.99

a)   What function is being described here?

b)   What is the domain and the range of the function described in part a)?

c)   How much would 9 CD's cost?

2) Suppose you are reading a newspaper article about crime in your local town and you happen upon the following figure.

Number of Car Thefts in Fairfield

a)   What is the function being described here?

b)   What is the domain and range of this function?

3) Let $f(x) = x^3 + 2x$. Determine:

       a) $f(2)$ b) $f(3)$ c) $f(x+h)$

       d) $f(2x)$ e) $f(-x)$ f) $f(2+\Delta x)$

4) Evaluate:

       a) $\left|3-4\right|$ b) $\left|4-3\right|$ c) $\left|2-18\right|$ d) $\left|\dfrac{3}{5}-\dfrac{4}{3}\right|$

5) Suppose $f(x) = x^4 - \sqrt{x}$, then determine:

       a) $f(0)$ b) $f(4)$ c) $f(x+h)$

       d) $f\left(\dfrac{\pi^2}{4}\right)$ e) $f(-x)$

6) Solve $\left|x\right| - 2 < 0$.   (Hint: Graph the function $f(x) = \left|x\right| - 2$, and see when it is strictly below the x-axis.)

7) Solve $x^2 - 9 > 0$.

3.3 <u>Lines and Their Equations</u>

Straight lines are the simplest of all curves, and one of the main ideas of calculus is to use lines to approximate complicated curves. So you've got to be proficient at lines and their equations. Lines have various degrees of steepness, which in mathematics is called <u>slope</u>. Consider Figure 3 below.

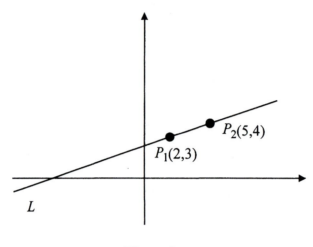

Figure 3

As you go from point P_1 to point P_2, you increase vertically a distance of 1 (called the <u>rise</u>), and horizontally a distance of 3 (called the <u>run</u>). The slope of line L is defined as $\dfrac{\text{rise}}{\text{run}} = \dfrac{1}{3}$.

Suppose the line L contains the two points $P_1(x_1, y_1)$ and $P_2(x_2, y_2)$ (see Figure 4 below),

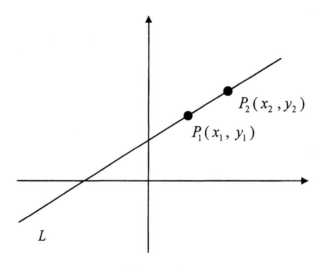

Figure 4

then L is said to have slope $m = \dfrac{y_2 - y_1}{x_2 - x_1}$, if $x_2 - x_1 \neq 0$.

<u>Remarks:</u>   1) Notice that if $x_2 - x_1 = 0$, then $x_2 = x_1$ and the line is vertical, and hence we have defined the notion of slope only for nonvertical lines.

2) Notice also what happens if we choose not P_1 and P_2, but any two other points. By using similarity of triangles we can prove that the slope calculated using any two points on L is equal to the slope obtained using P_1 and P_2.

We have talked about graphs of functions. What about graphs of equations? Here's the idea.

<u>Definition:</u>   The <u>graph of an equation</u> in x and y is the set of points (a,b) whose coordinates satisfy the equation when a is substituted for x and b is substituted for y.

Example 1:   Consider the line of slope -1 through the point $(-2,1)$. Find an equation whose graph is that line. Equivalently: What is the equation of the line of slope -1 through $(-2,1)$?

Solution: Consider the picture below:

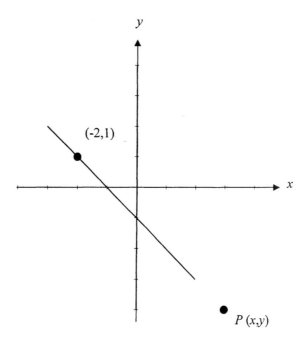

Does $P(x,y)$ lie on the line? It does if, and only if, the slope is correct, that is, if $\dfrac{\text{rise}}{\text{run}} = -1$, meaning

$$\frac{y-1}{x-(-2)} = -1.$$

We get rid of the denominator by multiplying by it, and then get

$$y - 1 = (-1)(x + 2).$$

So that is the equation of the line. ∎

<u>Theorem:</u>   The equation of the line of slope m going through the point (x_0, y_0) is

$$\boxed{y - y_0 = m(x - x_0).}$$

Proof: We'll use the same idea as in the last example.

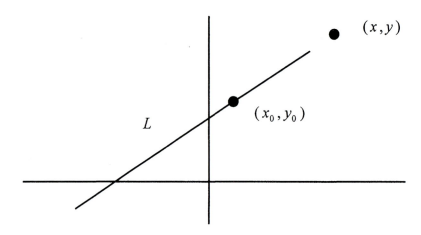

The point (x, y) lies on the line L if, and only if, the slope calculated from (x, y) and (x_0, y_0) is correct, that is: m. In other words $\dfrac{y - y_0}{x - x_0} = m$, which means $y - y_0 = m(x - x_0)$. (Notice that x and y are the variables, while m, x_0, and y_0 are just numbers.) This is called the <u>point-slope form</u> of the equation of a line. It is very important! Memorize it! (Right now is an excellent time.)

Example 2:   Find the equation of the line L of slope -3 going through the point $(2, -5)$.

Solution: Nothing to it if you've memorized that equation. In this case, $m = -3$, $(x_0, y_0) = (2, -5)$, and x and y are just the variables in the equation.

So $y - (-5) = -3(x - 2),$

which cleans up to be

$$y + 5 = -3(x - 2).$$

It's easy, once you know the slope and a point on the line. ∎

Example 3:   Consider the lines through the origin of slope $m = 1, 2, -3$. Find their equations and graph them.

Solution: We'll use the form $y - y_0 = m(x - x_0)$. In our case $(x_0, y_0) = (0,0)$, and so

$$y - 0 = m(x - 0),$$

which simplifies to

$$y = mx.$$

So the three given lines have equations $y = x$, $y = 2x$, and $y = -3x$. Their graphs are shown in Figures 5(a), 5(b), and 5(c) below, respectively. We can get the graphs by plotting merely one point in addition to the origin. Go 1 to the right, and up m, right? ■

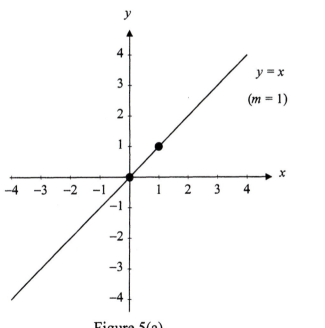

Figure 5(a)

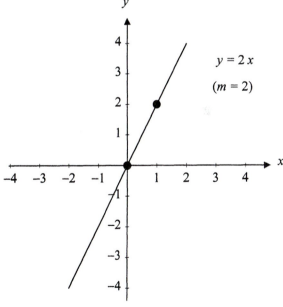

Figure 5(b)

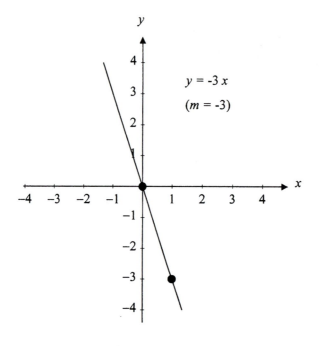

Figure 5(c)

Pictured below in Figure 6 is the <u>family</u> of functions of the form $y = mx$, for several m, all on the same set of axes for comparison purposes. As we go from left to right (our usual assumption), if m is positive, the line rises; if it is negative, the line sinks. The larger the absolute value of the slope, $|m|$, the steeper the line becomes. (What happens when the slope is a really big positive number?)

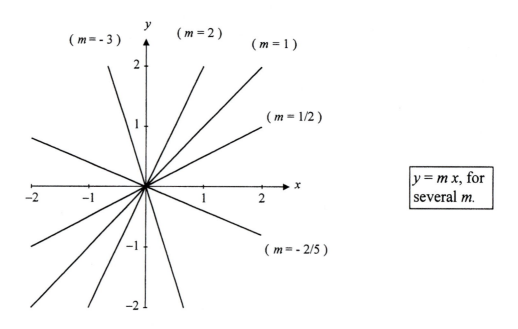

Figure 6

<u>Remark:</u>   It helps, whenever possible, to use the same scale on the *x*- and *y*-axes, meaning the distance from 0 to 1 is the same on the two axes. Sometimes that is really not feasible, as in exercise 5 below. In any event, always mark your scales – that is, show where 1, 2, etc. (or other appropriate values) are.

All non-vertical lines cross the *y*-axis. If a line crosses the *y*-axis at the point $(0,b)$, then *b* is called the <u>y-intercept</u>. To find the equation of the line of slope *m* and *y*-intercept *b* is easy. We use the point-slope form of the equation of a line, where the slope is *m*, and the line contains the point $(0,b)$. Doing so we get $y - b = m(x - 0)$ which reduces to

$$\boxed{y \ = \ mx + b.}$$

This is called the <u>slope-intercept form</u>. You should also memorize this.

Example 4:   Consider the equation $6x - 3y = 15$. What is its graph?

Solution: We could make a table by putting $x = 1, 2, 3$, etc., and calculating the corresponding *y*-values, and then plotting points, but we won't. Instead, we solve for *y*:

$$-3y = -6x + 15$$

or $$y = 2x - 5,$$

which we know is the equation of a line of slope 2 going through $(0,-5)$. If we start from $(0,-5)$ and go over 1 and up *m* (in this case $m = 2$), we get a second point on the line $(1,-3)$, and hence we know the whole graph.

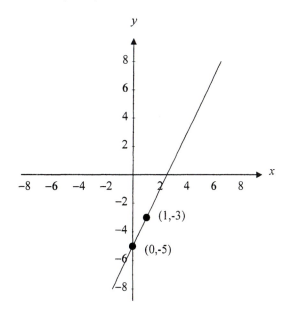

Lastly, we remind you that parallel lines have equal slopes, and perpendicular lines have slopes that are negative reciprocals of each other. This means, if line L_1 has slope m_1, and line L_2 has slope m_2, then, L_1 is parallel to L_2 if, and only if, $m_1 = m_2$, and L_1 is perpendicular to L_2 if, and only if, $m_2 = \dfrac{-1}{m_1}$.

Consider the following.

Example 5:   Find the equation of the line passing through the point $(1,3)$ that is parallel to the line $2x + 3y = 6$.

Solution: First, we determine the slope of the line given by putting it in slope-intercept form. We solve $2x + 3y = 6$ for y to get:

$$y = -\frac{2}{3}x + 2.$$

Comparing this to $y = mx + b$, we see that the slope of the given line is $-\dfrac{2}{3}$. Since the point on the line is $(1,3)$, we use the point-slope form to get

$$y - 3 = -\frac{2}{3}(x - 1),$$

which is one form of the answer. Depending on your purpose, you may wish to change it to slope-intercept form, or leave it as is. ■

Example 6:   Find the equation of the line passing through the point $(3,-2)$ that is perpendicular to the line $y = 6x + 4$.

Solution: The line $y = 6x + 4$ has slope equal to 6. Any line perpendicular to this line will have slope equal to $-\frac{1}{6}$. (This is the negative reciprocal of 6, right?) So the line we want passes through the point $(3,-2)$ and has slope $m = -\frac{1}{6}$. Using the point-slope form we obtain:

$$y - (-2) = -\frac{1}{6}(x - 3)$$

or

$$y = -\frac{1}{6}x - \frac{3}{2}. ■$$

Exercises 3.3

In Exercises 1–5, graph the function by plotting two points each. (Why is it enough to plot only two points ?)

1) $f(x) = \dfrac{1}{3}x$

2) $y = -1.5\,x$

3) $g(x) = .6\,x$

4) $f(x) = 5\,x$

5) $y = 1000\,x$

6) Find the equation of the line of slope $\dfrac{2}{3}$ through the point $(2,7)$.

7) Find the equation of the line of slope -6 through the point $(-3,2)$.

8) Find the equation of the line containing the two points $(1,2)$ and $(-2,1)$. (Hint: Find m.)

9) Find the equation of the line containing the two points $(-1,4)$ and $(2,1)$.

10) Graph the line given by the equation $4\,y - 3x = 24$.

11) Find the point-slope form for the equation of the line through the point $(1,4)$ parallel to the line $y = 2x - 5$. Graph the line.

12) Find the slope-intercept form for the equation of the line through the point $(-1,4)$ perpendicular to the line $y = \dfrac{1}{3}x - 5$. Graph this line, and label the y-intercept.

13) Find the equation of the line through the point $(\,0,\dfrac{4}{3})$ parallel to the line $y = 5x - 10$. Graph both lines.

14) Find the equation of the line through the point $(\dfrac{1}{2},-3)$ perpendicular to the line $y = -\dfrac{1}{7}x - 5$. Graph both lines.

3.4 Power Functions

Part A: The graphs of $y = x^r$, with $r = 2, 4$, and 6 are shown in Figure 7 below.

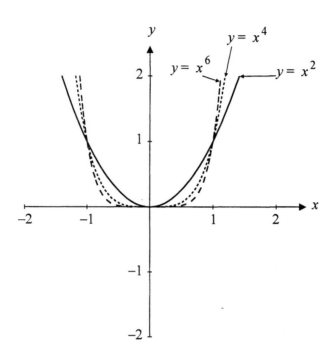

Figure 7

Notice how the points $x = 1$ and $x = -1$ are "pivotal?" Consider $x > 0$, or the right side of the graph. For $x < 1$, the relative position of the curves is opposite to what it is when $x > 1$. Well, let's think about it: for $x < 1$, the more we multiply it by itself the smaller it gets. The higher the power, the smaller the number gets. When $x > 1$, the opposite is true. The higher the power, the more we multiply by a number which is greater than 1, hence the larger the number gets. Notice that these graphs seem symmetric about the y-axis. In fact, they are. It doesn't matter whether we insert x or $-x$ into an even power function; the answer is the same. (Can you show this?) This shows symmetry about the y-axis.

Part B: The graphs of $y = x^r$, with $r = 3$, 5, and 7 are shown in Figure 8 below.

<u>Symmetry through the origin</u> is exhibited by the odd power functions.   Figure 4 shows three members of this little family.

As in Figure 7, the points $x = 1$ and $x = -1$ are pivotal in that the relative positioning of the graphs changes there. Now, however, the graphs are symmetric through the origin. That is, if the point (x, y) is on the graph, then so is $(-x, -y)$.

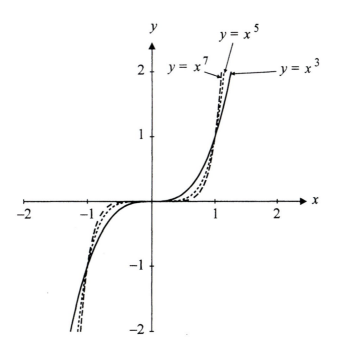

Figure 8

Part C: $y = x^r$, with $r = \frac{1}{2}, \frac{1}{4}, \frac{1}{6}$, etc., and $r = \frac{1}{3}, \frac{1}{5}, \frac{1}{7}$, etc.

What if r is a positive fraction of the form $\dfrac{1}{n}$ where n is a whole number 2, 3, 4, etc.? What does $y = x^{1/2}$, $y = x^{1/3}$, etc., look like? Well, as it turns out, this family of graphs also splits up into two groups – namely, the even n graphs and the odd n graphs. Within each group, the graphs display similar behavior. For $y = x^{1/2}$ and $x^{1/4}$, the graphs look like that in Figure 9. Since $y = x^{1/n}$ is equivalent to the nth root, if n is even x must not be negative.

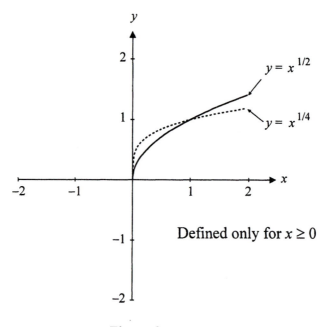

Figure 9

The other graphs of $y = x^{1/n}$ for n even are very similar to those shown in Figure 9, and as is the case in Figures 7 and 8, the point $x = 1$ is where the relative positioning of the graphs changes. If n is odd – say, for example, $y = x^{1/3}$ or $y = x^{1/5}$ – then x can be any real number. The graphs of these functions are shown in Figure 10.

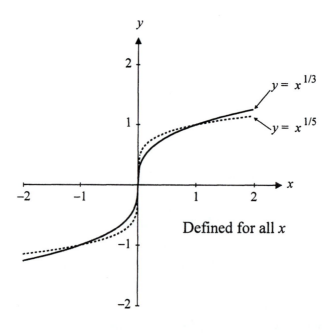

Figure 10

The other graphs of $y = x^{1/n}$ for n odd are very similar to those shown in Figure 10, with the same ordering situation.

To sum up Parts A, B, C: for $x \geq 0$, $y = x^r$ is shown in Figure 11.

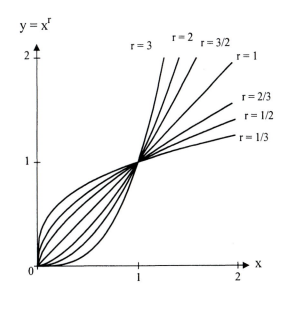

Figure 11

All of these are examples of <u>power functions</u>. Notice that $x^{2/3}$ is calculated by squaring $x^{1/3}$.

Part D: The graphs of $y = x^r$, with $r = -1$ and -3, are shown below in Figure 12, while the graphs of $y = x^r$, for $r = -2$ and -4, are shown in Figure 13 on the next page.

What if r is a negative integer? Can you guess that the situation will divide up into two cases, r even or odd? Here's the situation for r odd: $\dfrac{1}{x}, \dfrac{1}{x^3}$, etc. (Recall that x^{-3} means $\dfrac{1}{x^3}$, etc.)

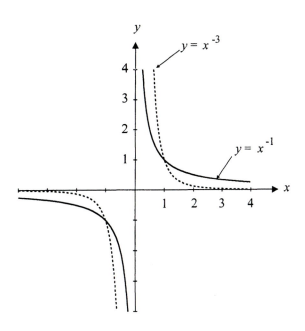

Figure 12

For $r = -2$ and -4, the graph is shown in Figure 13.

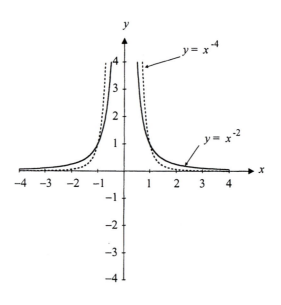

Figure 13

The functions x^{-3}, x^{-5}, etc., all basically look like x^{-1}, and x^{-4}, x^{-6}, etc., basically look like x^{-2}. The points $x = 1$ and $x = -1$ are again important since the relative positioning changes there.

Exercises 3.4   Using graph paper, calculate and plot accurately several points for each of the following:

1) $f(x) = \sqrt{x}$

2) $f(x) = x^3$

3) $g(x) = \sqrt[3]{x}$

4) $f(x) = \dfrac{1}{x}$

5) $g(x) = \dfrac{1}{x^2}$

6) $w(x) = x^{-3}$

7) $f(x) = x^{2/3}$

8) $f(x) = x^{3/2}$

3.5 **Shifting Up or Down**

How does the graph of $y = x^2 + 2$ compare to the graph of $y = x^2$? Well, all the y-coordinates of the first graph are 2 bigger than those of the second graph. But y is the altitude, or height, of the point (x,y). So to go from the graph of $y = x^2$ to $y = x^2 + 2$, just push up the graph a distance of 2, as shown in Figure 14.

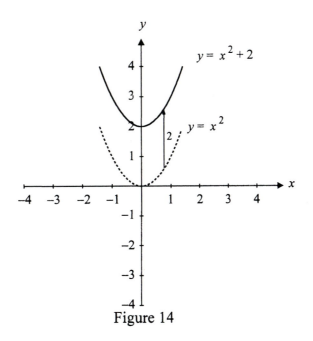

Figure 14

Also consider these examples:

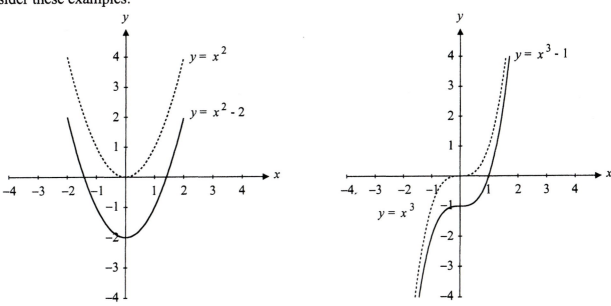

Figure 15(a)   Figure 15(b)

In each case, we are graphing a function by using the fact that it is shifted up or down from the graph of a function that we know.

Exercises 3.5

1) Graph the following functions:

a) $y = \dfrac{1}{2}x$ b) $y = \dfrac{1}{2}x + 2$ c) $y = \dfrac{1}{2}x - 1$

2) Graph the following functions:

a) $y = x^2 - 4$ b) $y = x^2 + \pi$ c) $y = x^{-2} + 1$

3) Graph the function $y = \sqrt{x} + 2$ for $x \geq 0$.

4) Graph the function $y = x^3 - 1$ for x in $[-2, 2]$.

5) Graph $|x|$ and $|x| - 2$.

3.6 <u>Shifting Left or Right</u>

Consider the following three graphs:

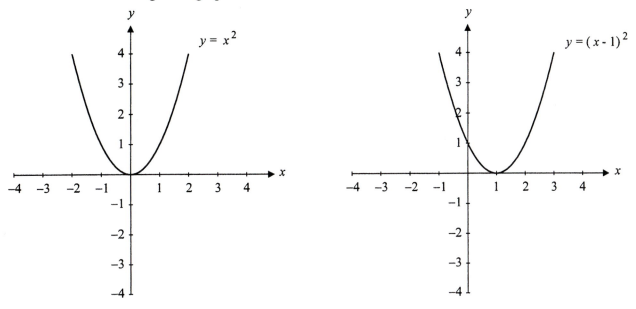

Figure 16(a)   Figure 16(b)

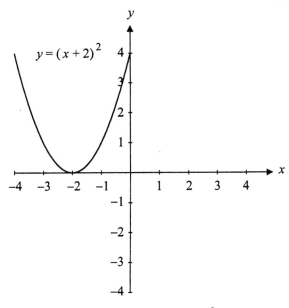

Figure 16(c)

(Don't take <u>our</u> word for it; test these by plotting points for $x = 0, \pm 1, \pm 2, \pm 3$.)

Notice that going from $f(x) = x^2$ to $f(x-1) = (x-1)^2$ we needed to shift the graph to the <u>right</u> a distance of 1, and going from $f(x) = x^2$ to $f(x+2) = (x+2)^2$ shifted the graph to the <u>left</u> a distance of 2. <u>This is true for all functions $f(x)$</u>, not just $f(x) = x^2$. In general, if we suppose a is some positive number, then the graph of $f(x-a)$ is the graph of $f(x)$ moved to the right a distance of a, and the graph of $f(x+a)$ is the graph of $f(x)$ moved to the left a distance of a. For instance, check the following graphs:

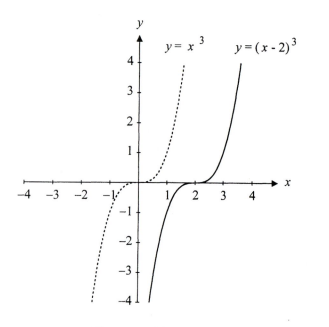

Figure 17(a)

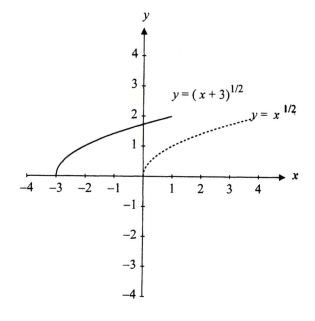

Figure 17(b)

Remarks: a) Notice in Figure 16(c) that the "vertex" of $(x + 2)^2$ is at -2, while in Figure 16(b) the vertex of $(x - 1)^2$ is at $+1$, which may be opposite to what you might have expected, but plotting a few points proves it.

b) It is important to see that, for example, $(x + 3)^2$ is very different from $x^2 + 3$. If we start from the graph of x^2, $(x + 3)^2$ moves it <u>3 units to the left</u>, while $x^2 + 3$ moves it <u>up a distance of 3 units</u>.

Exercises 3.6

1)   Graph x^2, $(x + 1)^2$, and $(x - 1)^2$ by plotting at least 5 points for each.

2)   Graph the following functions:

a) $f(x) = (x - 3)^2$ b) $y = (x + \pi)^2$ c) $y = \dfrac{1}{x-1}$

3)   Graph the following functions:

a) $y = (x + 3)^3$ b) $y = \sqrt[3]{x + 1}$ c) $y = -\dfrac{1}{(x - a)^4}$, where $a > 0$.

4)   Graph the function $y = \sqrt{x - 4}$ for $x \geq 4$.

5)   Suppose the dotted-line graph shown in the figure below $[\,g(x)\,]$ is obtained by shifting the solid line graph $[\,f(x)\,]$ down and over.   Express $g(x)$ in terms of $f(x)$.

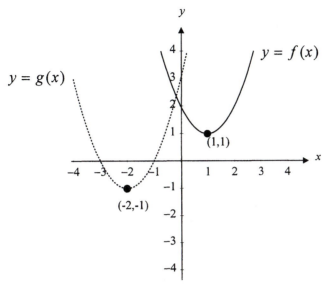

6)   Suppose you have the graph of a function $f(x)$.   Find a function $g(x)$ such that the graph of $g(x)$ is 2 to the right of, and 1 up from the graph of $f(x)$.

7)   Let $f(x)$ be a function.   Let h and k be numbers. Find a function $g(x)$ such that the graph of $g(x)$ is h units to the right of, and k units up from the graph of $f(x)$.

8)   Suppose $g(x)$ is a reflection of $f(x)$, and the graph of $k(x)$ is obtained from the graph of $g(x)$ by shifting right and down as shown.   Find an expression for $g(x)$ and $k(x)$.

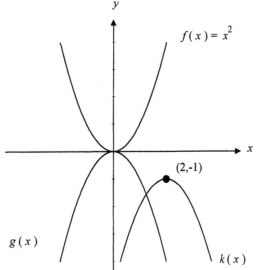

3.7 Intersection of Curves and Simultaneous Solutions

We know that a point (a,b) lies on a curve in the x-y plane if, and only if, $x = a$ and $y = b$ are solutions of the equation of that curve. (For example, the curve $y = x^2$ has the point $(2,4)$ on it because $2^2 = 4$.) What about a point of intersection of two curves? Well, that point lies on both curves, so its coordinates must satisfy the equations of both curves at the same time! That is, x and y must satisfy the two equations <u>simultaneously</u>. A set of two (or more) equations of two (or more) variables is called a <u>system of equations</u>. Finding solutions of such systems of equations is usually easy.

Example 1:   Find the intersection points of the curves whose equations are $y = x^2 - 4$ and $x + y = 8$.

Solution: Since both equations have to be true at the same time, we can solve for either x or y in one equation and substitute that result into the other equation. Here, let's solve for y in the first equation (it's already done!) and substitute that into the second equation, to give us

$$x + \left(x^2 - 4 \right) = 8.$$

We clean this up and put it into standard form:

$$x^2 + x - 12 = 0.$$

Here we can either use the quadratic formula or factor it to give

$$(x + 4)(x - 3) = 0.$$

By the Zero-Factor Property, $x = 3$ or -4. (See Section 2.3 for more on the ZFP.) So we know that there are two points of intersection: one where $x = 3$, and the other where $x = -4$. To find the corresponding y-values, you can substitute each x-value into either of the two equations (pick the easier one) to calculate y.

So: $y = x^2 - 4$

hence if $x = 3$, $y = 3^2 - 4 = 5$, and if $x = -4$, $y = (-4)^2 - 4 = 12$.
Therefore, the two points of intersection are $(3,5)$ and $(-4,12)$. ■

Example 2:   Where do the lines

$$2x + 3y = 7$$

and $$-3x + y = 11$$

intersect?

Solution: Here, let's start with the second equation to get $y = 3x + 11$, and substitute that into the first equation to get

$$2x + 3(3x + 11) = 7.$$

We clean this up and solve:

$$2x + 9x + 33 = 7$$

$$11x = -26$$

or $$x = \frac{-26}{11} = -\frac{26}{11}.$$

To find the corresponding y-value, use $y = 3x + 11$ with $x = -\frac{26}{11}$ in it to get

$$y = 3\left(-\frac{26}{11}\right) + 11$$

or $$y = -\frac{78}{11} + 11 = \frac{-78 + 121}{11} = \frac{43}{11}.$$

Hence there is one intersection point: $\left(-\frac{26}{11}, \frac{43}{11}\right)$. ∎

Remark: The method we used in these two examples is called the <u>method of substitution</u>. If all the equations are of degree 1 in whatever variables there are (like in Example 2), there is another way we can write the system using matrices. You can then use methods from the mathematical subject of linear algebra to easily solve such problems.

Exercises 3.7

1) Find the simultaneous solutions to the following systems of equations:

a) $\begin{cases} 3x - 2y = 16 \\ 5x + y = 5 \end{cases}$ b) $\begin{cases} x^2 - 4y = 6 \\ 2x + 2y = 3 \end{cases}$

2) Find the intersection of the curves:

a) $x^2 = y - 2$ and $2x + 3y = -7$

b) $x + y = 2$ and $\sqrt{x} + 4y = -6$

Chapter 4

Changing the Form of a Function

In most limit problems that you'll meet, there will be a fractional expression in which the numerator and denominator have a common factor that needs to be canceled out. Moral: you need to know how to factor to be a player in this game. It isn't always obvious how you should factor, but in your factoring toolbox you've got four basic methods that should be tried in this order:

1) Common factors

2) Special formulas

3) Grouping

4) The factor theorem

Recall that factoring an expression means writing it as a product.

4.1 Common Factors

The easiest tool to use is to take out all factors that are common to all the terms of the expression.

Example 1: Factor $3x^2y^3 + 15xy^4 - 21x^3y^2$.

Solution: Notice that all three terms share a factor of 3, as well as x and y^2. So, taking out this common factor of $3xy^2$, we get

$$3x^2y^3 + 15xy^4 - 21x^3y^2 = 3xy^2(xy + 5y^2 - 7x^2). \blacksquare$$

Remark: This is the easiest method of factoring and should always be done before going on to the rest.

Exercises 4.1 Factor the following expressions:

1) $2xy + 4x$

2) $6wz + 2wzt$

3) $xy + 4x + 2xw$

4) $6x^2y + 3xy + 9xy^2$

5) $10x^8y^6 + 25x^2y^4 + 20x^3y^{10}$

6) $24x^2yz + 2xy^2z^2 + 4xyz^3$

4.2 Special Formulas

(i) $\quad x^2 - y^2 = (x+y) \cdot (x-y)$

(ii) $\quad x^3 + y^3 = (x+y) \cdot (x^2 - xy + y^2)$

(iii) $\quad x^3 - y^3 = (x-y) \cdot (x^2 + xy + y^2)$

(iv) $\quad x^2 + (a+b)x + ab = (x+a) \cdot (x+b)$

(v) $\quad acx^2 + (bc+ad)x + bd = (ax+b) \cdot (cx+d)$

(vi) $\quad x^2 + 2xy + y^2 = (x+y)^2$ and $x^2 - 2xy + y^2 = (x-y)^2$

[The formulas of (vi) are special cases of the two previous formulas. They are called perfect squares.]

Remark: You should know formulas (i), (ii), (iii), and (vi) "in your sleep." The others, especially (v), are not so critical. All of them can be checked by multiplying out the right-hand sides.

Formula (i) is called the difference of two squares and comes up a lot. Consider the following examples.

Example 1:

a) $\quad z^2 - 9 = (z+3) \cdot (z-3)$ (Here we are using (i) with $x = z$ and $y = 3$.)

b) $\quad x^4 - y^2 = (x^2 + y) \cdot (x^2 - y)$ (Now x is replaced by x^2.)

c) $\quad (x-y)^2 - 4y^2 = (x-y+2y) \cdot (x-y-2y)$

$$= (x+y) \cdot (x-3y) \quad \blacksquare$$

Formula (ii) is called the sum of two cubes, while formula (iii) is the difference of two cubes. Notice how similar they are. One way to remember the signs in the formula is as follows: whatever sign is between the two cubes matches the sign between x and y in the first factor of the product, while the sign in front of the xy in the second factor is opposite to that. [Notice the $-xy$ in the middle of formula (ii) and the xy in the middle of formula (iii). It sometimes surprises people.]

Example 2:

a) $a^3 + 8b^3 = a^3 + (2b)^3 = (a+2b)\cdot(a^2 - 2ab + 4b^2)$

b) $27x^3 + 64y^3z^6 = (3x)^3 + (4yz^2)^3$

$$= (3x+4yz^2)\cdot(9x^2 - 12xyz^2 + 16y^2z^4)$$

Here we are using Formula (ii) with x replaced by $3x$ and y replaced by $4yz^2$.

c) $a^3 + 2b^3$ is the sum of the cubes of a and $\sqrt[3]{2}\,b$ (which equals $2^{\frac{1}{3}}b$). So,

$$a^3 + 2b^3 = (a + 2^{\frac{1}{3}}b)\cdot(a^2 - 2^{\frac{1}{3}}ab + 2^{\frac{2}{3}}b^2).\ \blacksquare$$

The usage of Formula (iii) goes pretty much the same way as that of (ii).

Example 3:

$$x^3 - 64y^6 = (x - 4y^2)\cdot(x^2 + 4xy^2 + 16y^4)$$

Again watch the sign of that middle term in the last factor! $\blacksquare$

Formula (iv) is used by noticing that the middle term is the <u>sum</u> of a and b, while the last term is their <u>product</u>.

Example 4:   Factor $x^2 + 5x + 6$.

Solution: We are looking for two numbers a and b so that their product is 6 and their sum is 5. Well, 6 factors only as 2 times 3, 6 times 1, −2 times −3, or −6 times −1. Of all these possibilities the only pair to add up to 5 is 2 and 3.

So: $x^2 + 5x + 6 = (x+2)\cdot(x+3).\ \blacksquare$

Example 5:   Factor $y^2 - 10y + 21$.

Solution: We need two numbers whose product is 21, so we think of 3 and 7, -3 and -7, 21 and 1, and -21 and -1. However, we notice that their sum must be -10, so we use -3 and -7.

$$y^2 - 10y + 21 = (y - 3) \cdot (y - 7)\ \blacksquare$$

Example 6:   Factor $s^2 + 2s - 8$.

Solution: Now we need two numbers whose product is -8 and sum is 2. The number -8 is the product of -8 and 1, -4 and 2, 4 and -2, or 8 and -1. The correct pair that adds up to 2 is 4 and -2, hence

$$s^2 + 2s - 8 = (s + 4) \cdot (s - 2).\ \blacksquare$$

Example 7:   Factor $a^4 - a^2 - 6$.

Solution: This looks different, but if we let $x = a^2$ we get $x^2 - x - 6$, and then this method of factoring applies. Now, -6 is the product of -6 and 1, -3 and 2, -2 and 3, and -1 and 6. Since their sum has to be -1, the only possible choice is -3 and 2. So,

$$x^2 - x - 6 = (x - 3) \cdot (x + 2)$$

and hence

$$a^4 - a^2 - 6 = (a^2 - 3) \cdot (a^2 + 2).$$

This can be factored further to give

$$a^4 - a^2 - 6 = (a - \sqrt{3})(a + \sqrt{3})(a^2 + 2).\ \blacksquare$$

Remark: In the last example the factorization to $(a^2 - 3) \cdot (a^2 + 2)$ is as far as you can go if you factor "over the integers," meaning that all the coefficients must be integers. If you factor "over the reals," you must go the extra step and get $(a - \sqrt{3})(a + \sqrt{3})(a^2 + 2)$. In some other contexts, you can even factor $a^2 + 2$ "over the complex numbers" into $(a + \sqrt{2}\,i)(a - \sqrt{2}\,i)$. In most situations in calculus, you'll need to factor only over the reals.

Example 8:   Factor $y^2 + 10y - 24$.

Solution: Notice how -24 is the product of many pairs, so this method of factoring can get to be tedious – in fact it can be a real pain! Since their sum is 10, the pair that works is 12 and -2, so

$$y^2 + 10y - 24 \; = \; (y-2)\cdot(y+12). \; \blacksquare$$

Ray of Hope: If this method is too tedious, or too tough, or if the factors are not whole numbers (which happens often), you can use a different method based on the quadratic formula and the factor theorem. For example, $x^2 - 2x - 4$ can't be factored easily without the factor theorem, which we will see in Section 4.4.

Formula (v) is usually really tough to use because too many possibilities need to be checked, unless you're lucky and hit the right one quickly. Your best bet in this case is to factor out the coefficient of the first term and work with the remaining expression. And there's always the factor theorem.

Exercises 4.2   Factor the following expressions. (Here, to factor means to factor as far as possible over the reals; see Example 7.)

1) $4y^2 - 9z^2$

2) $16x^4 - y^6$

3) $8s^3 + 27t^3$

4) $2x^3 + 64y^3$

5) $8s^3 - 27t^3$

6) $64z^3 - 9t^3$

7) $x^2 + 2x + 1$

8) $x^2 + 6x + 8$

9) $x^2 - 2x - 24$

10) $a^4 - 2a^2 - 24$

11) $s^6 - 7s^3 - 8$

12) $3x^2 + x - 2$

4.3 Grouping

This method sometimes, but not always, does the job. Here's how it works.

> Example 1:   Factor $10xy + 15y + 4x + 6$.
>
> Solution: As written, there are no common factors, but notice that the first and second terms have a common factor of $5y$, while the third and fourth terms have a common factor of 2, giving
>
> $$(10xy + 15y) + (4x + 6) = 5y(2x + 3) + 2(2x + 3).$$
>
> Now these two terms have a common factor of $2x + 3$, which we factor out to get
>
> $$10xy + 15y + 4x + 6 = (2x+3)\cdot(5y+2).$$
>
> Notice that the expression is in factored form. ■

> Example 2:   Factor $6ax + 3ay - 4bx - 2by + 10x + 5y$.
>
> Solution: You can group in more than one way. Let's try
>
> $$(6ax + 3ay) + (-4bx - 2by) + (10x + 5y)$$
>
> $$= 3a(2x + y) + (-2b)(2x + y) + 5(2x + y)$$
>
> $$= (2x + y)\cdot(3a - 2b + 5). ■$$

Remarks: a)   In the above example, we could have grouped differently. Noticing that some terms have an x-factor and some a y-factor, we could have done this:

$$6ax + 3ay - 4bx - 2by + 10x + 5y$$

$$= (6ax - 4bx + 10x) + (3ay - 2by + 5y)$$

$$= 2x(3a - 2b + 5) + y(3a - 2b + 5)$$

$$= (3a - 2b + 5)(2x + y)$$

The resulting factorization is the same as before. Always, if grouping is going to work, the resulting factorization will be the same regardless of how you group.

b) Not all expressions can be factored by grouping; in fact, some expressions cannot be factored <u>by any method</u>. If an expression has a prime number of terms, it definitely can't be factored by <u>grouping</u>. (Can you explain why?)

c) The resulting factors may need further factoring afterwards.

Exercises 4.3 Factor the following completely:

1) $3ax + 2ay + 3bx + 2by$

2) $x^4 - x^3y + x - y$

3) $x^{10} + x^6y^2 + x^4y^3 + y^5$

4) $6x^3y - 4xy^3 + 12yx^2 - 8y^3$

5) $3x^2 + 5xy + 7x + 3xy + 5y^2 + 7y$

4.4 **The Factor Theorem and Long Division**

Here it comes, the long-awaited entrance of the ultimate in theoretical sensations. Yes folks, the one, the only, the celebrated factor theorem. Here is the pearl of wisdom. (Let it rip Frank!)

Factor Theorem: Let $P(x)$ be a polynomial. Let a be any real number. Then $x - a$ is a factor of $P(x)$ if and only if $P(a) = 0$.

Remark: So what's the big deal with this factor theorem? Here's what. Suppose you want to factor a polynomial $P(x)$, and the other methods don't look promising. But suppose you can find some number a such that $P(a) = 0$. [You can do this by trying easy numbers like $a = \pm 1, \pm 2$, etc. to see if you are lucky and find such an a. If you have a graphing calculator, you can see where the graph of the function crosses or touches the x-axis. By zooming in on one such point, you may obtain a number a where $P(a) = 0$.] IF YOU CAN FIND an a with $P(a) = 0$, then you are sure, because of the factor theorem, that $x - a$ is a factor of $P(x)$.   So $P(x) = (x - a) \cdot (\text{something})$. That "something" can be found by long division, and so you've factored $P(x)$!

Example 1:   Factor $P(x) = x^3 - 2x^2 - 5x + 6$.

Solution: This one looks tough. The earlier methods of factoring would give you nothing. Zip. Zero. Nada. Goose eggs. But notice that if you put $x = 1$, you get $1^3 - 2 \cdot 1^2 - 5 \cdot 1 + 6 = 0$. Aha! So the F. T. (factor theorem) says that $x - 1$ is a factor. So:

$$P(x) = (x - 1) \cdot (\text{something}).$$

You're half done. To get the other factor, use long division:

$$x - 1 \overline{)x^3 - 2x^2 - 5x + 6}$$ (Notice that both expressions are in descending powers!)

Divide the first term into the first term (x into x^3 gives x^2). Put the x^2 on top, multiply it by the <u>entire divisor</u> $x - 1$, and subtract it from the dividend $x^3 - 2x^2 - 5x + 6$:

$$
\begin{array}{r}
x^2 \\
x - 1 \overline{)x^3 - 2x^2 - 5x + 6} \\
\underline{x^3 - x^2 } \\
-x^2 - 5x + 6
\end{array}
$$

(The easiest way to subtract $x^3 - x^2$ is to <u>mentally</u> change the sign of $x^3 - x^2$, getting $-x^3 + x^2$, and <u>adding</u>.) Continue in the same way: x into $-x^2$ gives $-x$, etc., to get:

$$
\begin{array}{r}
x^2 - x \phantom{{}- 5x + 6} \\
x-1 \overline{)\,x^3 - 2x^2 - 5x + 6} \\
\underline{x^3 - x^2} \\
-x^2 - 5x + 6 \\
\underline{-x^2 + x} \\
-6x + 6
\end{array}
$$

One more time: x into $-6x$ gives -6. So:

$$
\begin{array}{r}
x^2 - x - 6 \\
x-1 \,\overline{)\,x^3 - 2x^2 - 5x + 6} \\
\underline{x^3 - x^2} \\
-x^2 - 5x + 6 \\
\underline{-x^2 + x} \\
-6x + 6 \\
\underline{-6x + 6} \\
0
\end{array}
$$

We knew that the remainder would be 0 because the F. T. told us so.

Hence $P(x) = (x-1) \cdot (x^2 - x - 6)$.

So $P(x)$ is factored, but not factored completely as yet because $(x^2 - x - 6)$ can be factored further to give

$$P(x) = (x-1) \cdot (x-3) \cdot (x+2).$$

Success ! ■

Example 2:   Factor $2x^2 + 3x - 2$.

Solution: You could try special formula (v), but it would take too long. Remembering the F. T., all you need is a number "a" that makes $2a^2 + 3a - 2 = 0$. In ordinary English: you need a solution to $2x^2 + 3x - 2 = 0$. But that's easy, if you recall the quadratic formula,

$$x = \frac{-b \pm \sqrt{b^2 - 4ac}}{2a},$$

where in this context, $a = 2$, $b = 3$, and $c = -2$. Using this formula, you get

$$x = \frac{-3 \pm \sqrt{9 + 16}}{4} = \frac{-3 \pm 5}{4} = \frac{1}{2} \text{ or } -2.$$ So the factor theorem gives us both factors $\left(x - \frac{1}{2}\right)$ and $(x + 2)$, and so

$$2x^2 + 3x - 2 = \left(x - \frac{1}{2}\right) \cdot (x + 2) \cdot (\text{something}).$$

Since the coefficient of the x^2 term on the left is 2, that something <u>must be 2</u>, for equality. So:

$$2x^2 + 3x - 2 = \left(x - \frac{1}{2}\right)(x + 2)\,(2) = (2x - 1) \cdot (x + 2). \ \blacksquare$$

Example 3:   Can $x^2 + x + 1$ be factored?

Solution: The F. T. tells us to check for solutions of the equation $x^2 + x + 1 = 0$. The quadratic formula gives

$$x = \frac{-1 \pm \sqrt{1 - 4}}{2},$$

which is not a real number, because negative numbers don't have square roots among the real numbers. (See Section 1.5 for further explanation of this.) So there is no real number a to make $a^2 + a + 1$ equal to 0. Hence the F. T. says that $x^2 + x + 1$ <u>can't be factored</u>. (Such quadratics are called <u>irreducible</u>.) $\blacksquare$

To sum up: You can always use the F. T. to factor quadratics by using the quadratic formula, but for other polynomials $P(x)$, you'll need to be lucky to find a number a that makes $P(a) = 0$. Try $a = \pm 1, \pm 2$, etc.

In real-life problems, scientists and engineers use calculators or computers to find, or approximate, values for a where $P(a) = 0$.

Exercises 4.4   Factor the following expressions, if possible:

1) $x^3 - 3x + 2$

2) $2x^3 + x + 3$

3) $x^3 - x + 6$

4) $2x^2 - 3x + 4$

5) $x^2 - 3x - 2$

6) $24x^2 - 48x - 72$

4.5 Rationalizing Numerators or Denominators using Conjugates

Consider the expression $\dfrac{x - \sqrt{2}}{5}$. Notice that the numerator has two terms, one of which is a square root. For various reasons, you may not want a square root in the numerator. You can always get rid of it by multiplying and dividing by its conjugate $x + \sqrt{2}$. (You find the <u>conjugate</u> by changing the sign between the two terms.) So you will get

$$\frac{x - \sqrt{2}}{5} = \frac{x - \sqrt{2}}{5} \cdot \frac{x + \sqrt{2}}{x + \sqrt{2}} = \frac{(x - \sqrt{2}) \cdot (x + \sqrt{2})}{5 \cdot (x + \sqrt{2})}$$

$$= \frac{x^2 - 2}{5 \cdot (x + \sqrt{2})} \; .$$

Remarks: a) The top is now root-free.

b) The little devil's conjugate has popped up in the denominator instead. Depending on the problem you're solving, that may not cause any difficulties.

c) It all boils down to this: you can <u>exchange</u> a square root in the top for a square root in the bottom, or vice versa, whichever is better in a particular case.

Example 1: Rationalize the denominator of $\dfrac{x^2 - 3}{x + \sqrt{3}}$.

Solution: This means get rid of the root in the bottom. So multiply both top and bottom by the conjugate $x - \sqrt{3}$, giving

$$\frac{x^2 - 3}{x + \sqrt{3}} = \frac{x^2 - 3}{x + \sqrt{3}} \cdot \frac{x - \sqrt{3}}{x - \sqrt{3}} = \frac{(x^2 - 3)(x - \sqrt{3})}{(x + \sqrt{3})(x - \sqrt{3})}$$

$$= \frac{(x^2 - 3)(x - \sqrt{3})}{(x^2 - 3)} = (x - \sqrt{3}) \; . \; \blacksquare$$

Remark: When you're working with a fractional expression, and either the top or bottom has two terms, one (or both) of which is a square root, it is often useful to rationalize it. This comes up in limits, but also in many other cases. If you have such an expression, it <u>should always pop into your mind that one option is to rationalize</u>, which may help.

Example 2:   Rationalize $\dfrac{x^4 - 25}{x - \sqrt{5}}$.

Solution: Multiply both top and bottom by the conjugate $x + \sqrt{5}$, giving

$$\frac{x^4 - 25}{x - \sqrt{5}} = \frac{x^4 - 25}{x - \sqrt{5}} \cdot \frac{x + \sqrt{5}}{x + \sqrt{5}} = \frac{(x^4 - 25)(x + \sqrt{5})}{(x^2 - 5)},$$

which can be factored to

$$= \frac{(x^2 - 5)(x^2 + 5)(x + \sqrt{5})}{(x^2 - 5)} = (x^2 + 5)(x + \sqrt{5}),$$

which is probably easier to deal with. ∎

Remark: You may have noticed that you can get the same result by factoring the top completely and then canceling:

$$\frac{x^4 - 25}{x - \sqrt{5}} = \frac{(x^2 + 5)(x^2 - 5)}{x - \sqrt{5}} = \frac{(x^2 + 5)(x + \sqrt{5})(x - \sqrt{5})}{(x - \sqrt{5})} = (x^2 + 5)(x + \sqrt{5}).$$

That's true, but you've still got to know how to rationalize (see Exercises 5.1).

Example 3:   Rationalize $\dfrac{\sqrt{x+h} - \sqrt{x}}{h}$.

Solution: Multiply both top and bottom by the conjugate $\sqrt{x+h} + \sqrt{x}$, giving

$$\frac{\sqrt{x+h} - \sqrt{x}}{h} = \frac{\left(\sqrt{x+h} - \sqrt{x}\right)\left(\sqrt{x+h} + \sqrt{x}\right)}{h\left(\sqrt{x+h} + \sqrt{x}\right)},$$

which when simplified gives

$$\frac{\sqrt{x+h} - \sqrt{x}}{h} = \frac{x+h-x}{h\left(\sqrt{x+h} + \sqrt{x}\right)} = \frac{1}{\left(\sqrt{x+h} + \sqrt{x}\right)}. \quad \blacksquare$$

The following example presents part of the calculation needed to find the derivative of

$$f(x) = \frac{1}{\sqrt{x}}.$$

Example 4:   Write $\dfrac{1}{\sqrt{x+h}} - \dfrac{1}{\sqrt{x}}$ as one fraction, and rationalize the resulting numerator.

Solution: First, use a common denominator to get

$$\frac{\sqrt{x} - \sqrt{x+h}}{\sqrt{x+h}\ \sqrt{x}}.$$

Now multiply both top and bottom by the conjugate $\sqrt{x} + \sqrt{x+h}$.

$$\frac{\sqrt{x} - \sqrt{x+h}}{\sqrt{x+h}\ \sqrt{x}} = \frac{\sqrt{x} - \sqrt{x+h}}{\sqrt{x+h}\ \sqrt{x}} \cdot \frac{\sqrt{x} + \sqrt{x+h}}{\sqrt{x} + \sqrt{x+h}}$$

$$= \frac{x - (x+h)}{\sqrt{x+h}\ \sqrt{x}\left(\sqrt{x} + \sqrt{x+h}\right)}$$

$$= \frac{-h}{\sqrt{x+h}\ \sqrt{x}\left(\sqrt{x} + \sqrt{x+h}\right)} \quad \blacksquare$$

Exercises 4.5   In Exercises 1–7, rationalize the top or bottom, and simplify.

1) $\dfrac{7}{\sqrt{2} - 1}$ 2) $\dfrac{x + \sqrt{2}}{2}$ 3) $\dfrac{3}{x - \sqrt{7}}$ 4) $\dfrac{x + 1}{x + \sqrt{11}}$

5) $\dfrac{x^2 - 3}{x - \sqrt{3}}$ 6) $\dfrac{x^4 - 36}{x + \sqrt{6}}$ 7) $\dfrac{x^8 - 9}{x^2 + \sqrt{3}}$

8) Let $f(x) = \dfrac{1}{\sqrt{2x}}$. Calculate $\dfrac{f(x+h) - f(x)}{h}$ and simplify as in Example 4.

4.6 Extracting Factors from Radicals

Radicals can be difficult when computing things, so it usually pays to make them as simple as possible. Extracting factors from under the radical sign is one way of simplifying.

Example 1:   Simplify $\sqrt[3]{250\,x^4 y^3}$.

Solution: Since $(a\,b)^n = a^n \cdot b^n$ for all n (as long as both sides are defined!), we have $(a\,b)^{\frac{1}{3}} = a^{\frac{1}{3}} \cdot b^{\frac{1}{3}}$ and $\sqrt[3]{a \cdot b} = \sqrt[3]{a} \cdot \sqrt[3]{b}$. In this example you can write $250x^4\,y^3 = (125x^3 y^3)(2x)$. (We chose the first factor to be a <u>perfect cube</u>.)

So $\sqrt[3]{250\,x^4 y^3} = \sqrt[3]{125 x^3 y^3}\sqrt[3]{2\,x} = 5xy\sqrt[3]{2\,x}$. ■

Example 2:   Simplify the radical by extracting all that you can from $\sqrt{25 x^8\, y}$.

Solution: $\sqrt{25 x^8\, y} = \sqrt{(25 x^8)\,y}$

$= \sqrt{25 x^8}\sqrt{y}$

$= 5 x^4 \sqrt{y}$ ■

Remarks: a) Notice that in Example 1 as you "pulled" the factor $125x^3 y^3$ out of the radical, it "became" its cube root $5x\,y$. Similarly, as you "pulled" $25 x^8$ out of the root in Example 2, it "changed into" its square root $5 x^4$.

b) When working with square roots and other even-powered roots, you must remember that $\sqrt{a^2} = |a|$, not just a . The next example will illustrate.

Example 3:   Simplify $\sqrt{36 x\, y^2 z^3}$.

Solution: $\sqrt{36 x\, y^2 z^3} = \sqrt{(36\, y^2\, z^2)(x z)}$, where the first factor is a perfect square.

So,

$$\sqrt{36\,x\,y^2 z^3} = \sqrt{(36\,y^2\,z^2\,)}\sqrt{(x\,z)}$$

$$= (\sqrt{36}\sqrt{y^2}\sqrt{z^2}\,)\sqrt{x\,z}$$

$$= 6|y||z|\sqrt{x\,z}\,. \quad\blacksquare$$

Example 4:   Simplify $\sqrt[4]{16\,x^8 y^3}$.

Solution: $\sqrt[4]{16\,x^8 y^3}$

$$= \sqrt[4]{16\,x^8}\,\sqrt[4]{y^3}$$

$$= \sqrt[4]{16}\,\sqrt[4]{x^8}\,\sqrt[4]{y^3}$$

$$= 2\,\left|x^2\right|\sqrt[4]{y^3}$$

Here, the stuff in the absolute value sign, x^2 , is always nonnegative, and so $\left|x^2\right| = x^2$. So the given expression equals

$$2\,x^2\sqrt[4]{y^3}\,. \quad\blacksquare$$

Example 5:   Simplify $\sqrt[5]{32\,x^{10}y^2}$.

Solution: $\sqrt[5]{32\,x^{10}}\,\sqrt[5]{y^2} = 2\,x^2\sqrt[5]{y^2} \quad\blacksquare$

Example 6:   Simplify $\sqrt{x^2 y^6 + 3\,x^5 y^4}$.

Solution: The stuff in the radical is not factored yet, so you must do that before you can extract any factors!

$$\sqrt{x^2 y^6 + 3\,x^5 y^4} = \sqrt{x^2 y^4 (y^2 + 3\,x^3\,)}$$

$$= \sqrt{x^2 y^4} \; \sqrt{y^2 + 3x^3}$$

$$= |x| |y^2| \sqrt{y^2 + 3x^3}$$

$$= |x| y^2 \sqrt{y^2 + 3x^3} \quad \blacksquare$$

Exercises 4.6   Extract as much as you can from the following roots:

1) $\sqrt{16 x^2}$

2) $\sqrt{4 x^2 + 8 x^4}$

3) $\sqrt[3]{54 x^4}$

4) $\sqrt{3 x^{12} y}$

5) $\sqrt{5 x^4 + 3 x^8}$

6) $\sqrt[3]{27 x^6 y}$

7) $\sqrt{8 \pi^2 x^3 y^4}$

8) $\sqrt[4]{x^5 y^4 + x^6 y^{10}}$

Chapter 5

Simplifying Algebraic Expressions

5.1 <u>Working with Difference Quotients</u>

At this point in calculus you're going to be meeting up with expressions called difference quotients. Such an expression looks like

$$\frac{f(x+h)-f(x)}{h}.$$

Alternatively, the difference quotient can be written using the variable Δx instead of the variable h. Hence you may be seeing things like

$$\frac{f(x+\Delta x)-f(x)}{\Delta x}.$$

Your ability to work with difference quotients often depends on your ability to expand expressions and cancel common factors. Some of the following examples feature difference quotients, while others are for general practice.

Example 1:   Let $f(x)=x^2$, calculate $\dfrac{f(x+h)-f(x)}{h}$.

Solution: Since $f(x+h)=(x+h)^2$,

$$\frac{f(x+h)-f(x)}{h}=\frac{(x+h)^2-x^2}{h}.$$

Expanding $(x+h)^2$ as $x^2+2xh+h^2$ we get

$$\frac{f(x+h)-f(x)}{h}=\frac{(x^2+2xh+h^2)-x^2}{h},$$

which simplifies to

$$\frac{f(x+h)-f(x)}{h}=\frac{2xh+h^2}{h}=\frac{h(2x+h)}{h}$$

$$=2x+h.\ \blacksquare$$

Remark: The last step in Example 1 was to cancel a common factor of h from the numerator and denominator. Notice that we are able to do this since <u>every</u> term of the numerator contained a factor of h.

Example 2:   Let $f(x) = 2x^2 - 3x$. Calculate $\dfrac{f(x + \Delta x) - f(x)}{\Delta x}$.

Solution: Since $f(x + \Delta x) = 2(x + \Delta x)^2 - 3(x + \Delta x)$,

$$\frac{f(x + \Delta x) - f(x)}{\Delta x} = \frac{2(x + \Delta x)^2 - 3(x + \Delta x) - \left(2x^2 - 3x\right)}{\Delta x}.$$

Expanding $(x + \Delta x)^2$ as $x^2 + 2x\Delta x + (\Delta x)^2$, we get

$$\frac{f(x + \Delta x) - f(x)}{\Delta x}$$

$$= \frac{2x^2 + 4x\Delta x + 2(\Delta x)^2 - 3x - 3\Delta x - 2x^2 + 3x}{\Delta x},$$

which simplifies to

$$\frac{f(x + \Delta x) - f(x)}{\Delta x} = \frac{4x\Delta x + 2(\Delta x)^2 - 3\Delta x}{\Delta x} = \frac{\Delta x(4x + 2\Delta x - 3)}{\Delta x},$$

$$= 4x + 2\Delta x - 3. \ \blacksquare$$

Computing difference quotients for functions other than polynomials gets a little tricky. An ability to add and subtract rational expressions becomes essential. Recall from Chapter 1 how to add fractions: $\dfrac{a}{b} + \dfrac{c}{d} = \dfrac{ad + bc}{bd}$. (First form a common denominator, namely bd, then change $\dfrac{a}{b}$ to $\dfrac{ad}{bd}$, and $\dfrac{c}{d}$ to $\dfrac{bc}{bd}$, at which point you can add them to get the result.) The same method applies, no matter how tough these expressions look.

Example 3:   Simplify $\dfrac{1}{x} - \dfrac{1}{x-1}$.

Solution: $\dfrac{1}{x} - \dfrac{1}{x-1} = \dfrac{1 \cdot (x - 1) - 1 \cdot x}{x(x - 1)}$

$$= \frac{x - 1 - x}{x(x - 1)} = \frac{-1}{x(x - 1)} \ \blacksquare$$

Example 4:   Simplify $\dfrac{3x + y}{x + y} + \dfrac{x - y}{x + 2y}$.

Solution: Finding a common denominator and adding gives

$$\frac{(3x + y)(x + 2y) + (x + y)(x - y)}{(x + y)(x + 2y)} .$$

This needs a little "cleaning up." Let's multiply out the top, to get

$$\frac{3x^2 + xy + 6xy + 2y^2 + x^2 + xy - xy - y^2}{(x + y)(x + 2y)}$$

$$= \frac{4x^2 + 7xy + y^2}{(x + y)(x + 2y)} .$$

For most purposes, you would not want to multiply out the denominator.   ■

Example 5:   Simplify $\dfrac{\dfrac{1}{t + 1} - \dfrac{1}{t}}{\dfrac{1}{t + 1} + \dfrac{1}{t}}$.

Solution: Working separately with the numerator and denominator,

$$\frac{1}{t + 1} - \frac{1}{t} = \frac{t - (t + 1)}{t(t + 1)} = \frac{-1}{t(t + 1)}$$

$$\frac{1}{t + 1} + \frac{1}{t} = \frac{t + (t + 1)}{t(t + 1)} = \frac{2t + 1}{t(t + 1)} .$$

So:

$$\frac{\dfrac{1}{t + 1} - \dfrac{1}{t}}{\dfrac{1}{t + 1} + \dfrac{1}{t}} = \frac{\dfrac{-1}{t(t + 1)}}{\dfrac{2t + 1}{t(t + 1)}} = \frac{-1}{t(t + 1)} \cdot \frac{t(t + 1)}{2t + 1} = \frac{-1}{2t + 1} .$$

This answer is as simple as it gets.   ■

Example 6: Simplify $\dfrac{x^{-1} + y^{-1}}{(xy)^{-1}}$.

Solution: First convert the expression to fractions:

$$\frac{x^{-1} + y^{-1}}{(xy)^{-1}} = \frac{\dfrac{1}{x} + \dfrac{1}{y}}{\dfrac{1}{xy}}$$

$$= \frac{\dfrac{1 \cdot y + 1 \cdot x}{xy}}{\dfrac{1}{xy}} = \frac{\dfrac{y + x}{xy}}{\dfrac{1}{xy}} = \frac{y + x}{1} = y + x. \quad \blacksquare$$

Example 7: Simplify $\dfrac{\dfrac{1}{x + h} - \dfrac{1}{x}}{h}$.

Solution:

$$\frac{\dfrac{1}{x + h} - \dfrac{1}{x}}{h} = \frac{\dfrac{x - (x + h)}{x(x + h)}}{h}$$

$$= \frac{\dfrac{x - x - h}{x(x + h)}}{h} = \frac{\dfrac{-h}{x(x + h)}}{h}$$

Notice that both the top and bottom have a factor of h, which can be canceled to obtain

$$\frac{h\left(\dfrac{-1}{x(x + h)}\right)}{h} = \frac{-1}{x(x + h)}. \quad \blacksquare$$

Example 8: Simplify $\dfrac{3x^2 y^5}{24 x y^2}$.

Solution: Cancel factors that are common to the numerator and the denominator:

$$\frac{3x^2 y^5}{24 x y^2} = \frac{x y^3}{8}. \quad \blacksquare$$

Example 9: Simplify $\dfrac{5a^2 b^3 - 15 a b}{100 a^2 b^2}$.

Solution: Remember to avoid "creative canceling." Cancel only those things that are factors of both the <u>entire top</u> and the <u>entire bottom</u>. Factor the numerator to obtain

$$\frac{5a^2 b^3 - 15 a b}{100 a^2 b^2} = \frac{5 a b(a b^2 - 3)}{100 a^2 b^2}.$$

Notice that the numerator and denominator both have a factor of $5 a b$.
Cancel it to get

$$\frac{a b^2 - 3}{20 a b}. \quad \blacksquare$$

Example 10: Simplify $\dfrac{27 x^2 \left(\dfrac{y}{z}\right)^{-5}}{(3 x y z^4)^2}$.

Solution: $\dfrac{27 x^2 \left(\dfrac{y}{z}\right)^{-5}}{(3 x y z^4)^2} = \dfrac{27 x^2 \left(\dfrac{z}{y}\right)^{5}}{9 x^2 y^2 z^8} = \dfrac{27 x^2 y^{-5} z^5}{9 x^2 y^2 z^8}$

$$= 3 y^{-7} z^{-3} = \frac{3}{y^7 z^3} \quad \blacksquare$$

Example 11: Let $f(x) = x^2 + \dfrac{1}{x}$.

Calculate: a) $f(x + \Delta x)$

b) $\dfrac{f(x + \Delta x) - f(x)}{\Delta x}$ (You'll see lots of these !)

Note: The symbol Δx is to be interpreted as just another variable that could just as well be called h (which it often is), or y or w or whatever.

Solution: a) $f(x + \Delta x) = (x + \Delta x)^2 + \dfrac{1}{x + \Delta x}$

b) $\dfrac{f(x + \Delta x) - f(x)}{\Delta x} = \dfrac{(x + \Delta x)^2 + \dfrac{1}{x + \Delta x} - \left(x^2 + \dfrac{1}{x}\right)}{\Delta x}$

$= \dfrac{x^2 + 2x \cdot \Delta x + (\Delta x)^2 + \dfrac{1}{x + \Delta x} - x^2 - \dfrac{1}{x}}{\Delta x}$

Cancel the x^2's, and subtract the two fractions on top. So

$\dfrac{f(x + \Delta x) - f(x)}{\Delta x} = \dfrac{2x \cdot \Delta x + (\Delta x)^2 + \dfrac{x - (x + \Delta x)}{x(x + \Delta x)}}{\Delta x}$

$= \dfrac{2x \cdot \Delta x + (\Delta x)^2 - \dfrac{\Delta x}{x(x + \Delta x)}}{\Delta x}$

$= \dfrac{\Delta x \left(2x + \Delta x - \dfrac{1}{x(x + \Delta x)}\right)}{\Delta x}$

$= 2x + \Delta x - \dfrac{1}{x(x + \Delta x)}$. ∎

Exercises 5.1

1) Let $f(x) = x^2 + 3x$.

   a) Compute $f(x+h)$.

   b) Simplify $\dfrac{f(x+h) - f(x)}{h}$.

2) Let $g(x) = 2x^3 - x$.

   a) Compute $g(x+h)$.

   b) Simplify $\dfrac{g(x+h) - g(x)}{h}$.

Simplify the expressions in Exercises 3–11.

3) $\dfrac{1}{x-1} - \dfrac{1}{x}$

4) $\dfrac{x}{x-1} - \dfrac{2}{x}$

5) $\dfrac{x}{x-1} - \dfrac{x}{x+1}$

6) $\dfrac{\dfrac{s+1}{s-1} + \dfrac{s-1}{s+1}}{\dfrac{1}{s^2-1}}$

7) $\dfrac{2(x+h) + 1 - (2x+1)}{h}$

8) $\dfrac{\dfrac{1}{(x+h)^2} - \dfrac{1}{x^2}}{h}$

9) $x + \dfrac{1}{x + \dfrac{1}{x + \dfrac{1}{x}}}$

10) $\dfrac{\sqrt{2x+2h} - \sqrt{2x}}{h}$ (Hint: Rationalize your way out of this one!)

11) $\dfrac{3a^2 b - 27 a b^2}{(15 a^3 b^4)^2}$

12)   Let $g(x) = \sqrt{x-3}$. Calculate $\dfrac{g(x+\Delta x) - g(x)}{\Delta x}$.   (Hint: Rationalize your way to success!)

Chapter 6

Composition and Decomposition of Functions

6.1 Composition

Consider the function $k(x) = \dfrac{1}{\sqrt{x}}$. Then

$$k(1) = \frac{1}{\sqrt{1}} = 1,$$

$$k(4) = \frac{1}{\sqrt{4}} = \frac{1}{2},$$

$$k(x+2) = \frac{1}{\sqrt{x+2}},$$

$$k(x^4) = \frac{1}{\sqrt{x^4}} = \frac{1}{x^2},$$

$$k(x^2+x+1) = \frac{1}{\sqrt{x^2+x+1}},$$

$$k(x+h) = \frac{1}{\sqrt{x+h}},$$

and
$$k(\text{whatever}) = \frac{1}{\sqrt{\text{whatever}}}.$$

Get it? Got it? Good!

Now consider the functions $f(x) = \sqrt{x}$ and $g(x) = x + 4$. Then $f\big(g(x)\big) = \sqrt{g(x)} = \sqrt{x+4}$. (Agreed?) Why are we looking at functions of functions? Because it happens all the time out there, in "real life." For example, the cost of gasoline for a trip is a function of the amount of gasoline you use, and the amount of gasoline you use is a function of your speed on the trip. Hence your gasoline cost is a function of a function.

Example 1:   Let $f(x) = x^2 + 2x + 3$ and $g(x) = x^3$. Then what is $f(g(x))$?

Solution: $f(g(x)) = (g(x))^2 + 2(g(x)) + 3$

$$= (x^3)^2 + 2x^3 + 3$$

$$= x^6 + 2x^3 + 3 \quad \blacksquare$$

Remarks: 1) It is important to emphasize that the composition $f(g(x))$ is not the same as the product $(f(x))(g(x))$, which in this case is $x^5 + 2x^4 + 3x^3$.

2) Notice that, in general, $f(g(x))$ is also not the same as $g(f(x))$, as the next example shows.

Example 2:   If $f(x) = x^4$ and $g(x) = \sqrt{x+1}$, then what is $f(g(x))$ and $g(f(x))$?

Solution: a) $f(g(x)) = (g(x))^4 = (\sqrt{x+1})^4 = ((\sqrt{x+1})^2)^2$

$$= (x + 1)^2 = x^2 + 2x + 1$$

b) $g(f(x)) = \sqrt{f(x) + 1} = \sqrt{x^4 + 1}$

Notice first that $f(g(x)) \neq g(f(x))$. Secondly, in part (b) you cannot simplify the radical $\sqrt{x^4 + 1}$ any further; $\sqrt{x^4 + 1}$ is not equal to $x^2 + 1$. $\blacksquare$

Definition: $f(g(x))$ is called the <u>composition of f with g</u>. There's a special notation for it, $f \circ g$, and its evaluation at a point x is denoted $(f \circ g)(x)$.

Example 3: Let $f(x) = \dfrac{1}{x+2}$ and $g(x) = x^2 - 1$. Find $(f \circ g)(x)$ and $(g \circ f)(x)$.

Solution: a) $(f \circ g)(x) = f\big(g(x)\big) = \dfrac{1}{g(x)+2} = \dfrac{1}{x^2 - 1 + 2} = \dfrac{1}{x^2 + 1}$

b) $(g \circ f)(x) = \big(f(x)\big)^2 - 1 = \left(\dfrac{1}{x+2}\right)^2 - 1$

Notice that in this example also, $f \circ g \neq g \circ f$. ∎

Does it ever happen that $f \circ g = g \circ f$? Sure, just try $f(x) = x + 2$ and $g(x) = x + 3$. Then, $f\big(g(x)\big) = (x + 3) + 2 = x + 5$, and $g\big(f(x)\big) = (x + 2) + 3 = x + 5$. An important special case occurs when $f\big(g(x)\big)$ and $g(f(x))$ are not only equal, but, in fact, are both equal to x. In that case, f and g are said to be <u>inverses</u> of each other. You will find more information on inverses in Chapter 9, but for now look at the next example.

Example 4: Let $f(x) = 3x$ and $g(x) = \dfrac{1}{3}x$. Find $(f \circ g)(x)$ and $(g \circ f)(x)$.

Solution: a) $(f \circ g)(x) = f\big(g(x)\big) = 3\left(\dfrac{1}{3}x\right) = x$

b) $(g \circ f)(x) = \dfrac{1}{3}\big(3x\big) = x$

Notice that $(f \circ g)(x) = (g \circ f)(x) = x$ in this case. One function "undoes" the action of the other function. This shouldn't surprise you since f triples a number and g divides the number by 3. These two arithmetic operations are called inverses of one another. ∎

We can define $f \circ g \circ h$ in a similar fashion as shown in the following example.

Example 5:   Let $f(x) = x^2$, $g(x) = x - 3$, and $h(x) = 2x + 1$. What is the composition $(f \circ g \circ h)(x)$?

Solution: $(f \circ g \circ h)(x) = f\big(g(h(x))\big)$

$$= f\big(g(2x+1)\big) = f\big((2x + 1) - 3\big) = f(2x-2)$$

$$= (2x - 2)^2 \quad \blacksquare$$

Example 6:   Let $f(x) = x^5 - 3x$ and $g(t) = t^2$. Find the function $(f \circ g)(t)$.

Solution: Notice here that g is a function of t, not x, but that's no problem. You simply substitute $g(t)$ into $f(x)$ to get

$$(f \circ g)(t) = f\big(g(t)\big) = f(t^2) = (t^2)^5 - 3(t^2) = t^{10} - 3t^2. \quad \blacksquare$$

Exercises 6.1

1) Suppose $f(x) = x^2 + 1$ and $g(x) = \sqrt{x}$, then what is the composition $(f \circ g)(x)$?

2) Suppose $f(x) = \dfrac{1}{x + 5}$ and $g(x) = x^3 + 2x - 3$, then what is the composition $(f \circ g)(x)$?

3) Given the functions

$$f(x) = x^3, \qquad g(x) = \sqrt{x + 1}, \qquad s(t) = 2t - 3, \qquad h(x) = x^2 - \frac{1}{x},$$

find the following composition functions:

   a) $(f \circ g)(x)$ b) $(f \circ s)(t)$ c) $(f \circ h)(x)$ d) $(g \circ f)(x)$

   e) $(g \circ s)(t)$ f) $(g \circ h)(x)$ g) $(s \circ f)(x)$ h) $(s \circ g)(x)$

   i) $(s \circ h)(x)$ j) $(h \circ f)(x)$ k) $(h \circ g)(x)$ l) $(h \circ s)(t)$

4) Suppose that $f(x) = x^2 - 2x$, $g(x) = \sqrt{x}$, and $h(x) = \dfrac{1}{x + 1}$. Find:

   a) $(f \circ g \circ h)(x)$ b) $(f \circ h \circ g)(x)$ c) $(g \circ h \circ f)(x)$ d) $(g \circ f \circ h)(x)$

6.2 Decomposition

In calculus, you will be taking derivatives of functions. If a function $k(x)$ is a composition of two or more functions, it is called a composite function (or simply a composite), and its derivative is found using the "chain rule." You will first need to decompose $k(x)$, which means you will need to first find functions f and g such that $k(x) = f(g(x))$; here f is called the outer function, and g is called the inner function. If this can be done in more than one way, be sure to let f be the outermost function.

Example 1:   Decompose the function $y(x) = (x^2 + 1)^5$.

Solution: If you were to evaluate this function (at $x = 1$ for example), here's the sequence of arithmetic steps you would take. First you would take $x = 1$ and square it (giving you 1), then add 1 (giving you 2), then take the result and raise it to the fifth power (giving you 32). The last operation would be to take the fifth (in a manner of speaking).

So $f(x) = x^5$ is the outermost function. Clearly $g(x) = x^2 + 1$ is the inner function and $(f \circ g)(x) = (g(x))^5 = (x^2 + 1)^5$.

Hence $f(x) = x^5$ and $g(x) = x^2 + 1$ gives the desired decomposition. ■

Example 2:   Decompose the function $y(x) = \sqrt{x^2 + x + 3}$.

Solution: Again, you should think about the order of steps this particular function requires. At $x = 2$, for example, first you would square 2 (giving you 4), then add 2 (giving you 6), then add 3 (giving you 9), and finally take the square root (giving you 3). Here, the last operation would be to take the square root, so $f(x) = \sqrt{x}$ is the outermost function and $g(x) = x^2 + x + 3$ is the inner function. ■

Example 3:   Decompose the function $y(x) = \left(x^2 + 1\right)^{5/3}$.

Solution: The outermost function is now $f(x) = x^{5/3}$, and now the inner function is $g(x) = x^2 + 1$. ∎

If you've already studied the functions e^x and $\ln x$, consider the next three examples.

Example 4:   Decompose the function $h(x) = \ln\left(x^2 + 1\right)$.

Solution: To find the outermost function, evaluate $h(x)$ at $x = 2$, for example. Inside the bracket you get 5; then you apply the $\ln x$ function to get $\ln 5$. So: the last operation was to apply $\ln x$, and hence $f(x) = \ln x$ is the outermost function, and $g(x) = x^2 + 1$ is the inner function. ∎

Example 5:   Decompose $e^{\sqrt{x+5}}$.

Solution: If you evaluate this at $x = 4$, for example, you get a 9 inside the root, and so $\sqrt{x + 5} = 3$. The last step is to apply the exponential function e^x to get e^3. Hence the outermost function is $f(x) = e^x$, and the inner function is $g(x) = \sqrt{x + 5}$. ∎

Example 6:   Decompose $\left(\ln x + e^x\right)^5$.

Solution: The outermost function is now $f(x) = x^5$, and now the inner function is $g(x) = \ln x + e^x$. ∎

Exercises 6.2

1) Find decompositions for the following functions – that is, find f and g such that $y = f \circ g$, and f is the outermost function.

    a) $\ y(x) = \sqrt{x+1}$ b) $\ y(x) = (x^3 - 1)^2$ c) $\ y(x) = (x^{1/2} + x)^7$

    d) $\ y(x) = \sqrt[3]{x^{2/3} + 2}$ ė) $\ y(x) = \left(x^5 + 3x^2 + x\right)^{-3/2}$

2) If you've already studied the functions e^x and $\ln x$, decompose these functions:

    a) $\ \ln\left(x^2 + 2\right)$ b) $\ e^{\sqrt{x} + 1}$ c) $\ \left(\ln x + e^x\right)^2$

    d) $\ \ln\left(x + e^x\right)$ e) $\ \sqrt{\ln\left(x^2 + 1\right)}$

Chapter 7

Equations of Degree 1 Revisited

7.1 Solving Linear Equations Involving Derivatives

You're asking: why are we doing this again? Well, in calculus, when you are using the method of implicit differentiation to compute the derivative of a quantity, what sometimes results is a rather nasty looking equation that contains lots of x's and y's and the derivative $\dfrac{dy}{dx}$ (or y', depending on your notation). However, if you have carried out the differentiation correctly, you will notice that in any terms containing $\dfrac{dy}{dx}$ it appears only to the first power. That is, there are no terms that contain $\left(\dfrac{dy}{dx}\right)^2$, or $\left(\dfrac{dy}{dx}\right)^3$, or even $\sqrt{\dfrac{dy}{dx}}$. This means that your equation is really a linear equation in the variable $\dfrac{dy}{dx}$. All the methods shown in Section 2.1 can be used here. Just remember that $\dfrac{dy}{dx}$ (or y') is just a symbol. It's a variable like x, y, z, t, or w. Treat it like one.

Example 1:   Solve for $\dfrac{dy}{dx}$: $2 + 4\dfrac{dy}{dx} = \dfrac{dy}{dx} + 1$.

Solution: Isolate the variable we wish to solve for, which in this case is the derivative $\dfrac{dy}{dx}$. Bring all the terms with $\dfrac{dy}{dx}$ to the left by subtracting $\dfrac{dy}{dx}$ from both sides to give

$$2 + 3\dfrac{dy}{dx} = 1.$$

Now subtract 2 from each side of the equation to give

$$3\dfrac{dy}{dx} = -1.$$

Lastly we divide by 3 to get the solution

$$\dfrac{dy}{dx} = -\dfrac{1}{3}. \quad \blacksquare$$

Example 2: Solve for y': $2x + 3y' = 3x - 5y'$.

Solution: Isolate the variable we wish to solve for, which in this case is the derivative y', by bringing all the terms with y' to the left, and all the others to the right. Do this by first adding $5y'$ to both sides of the equation. This gives us

$$2x + 8y' = 3x \ .$$

Now subtract $2x$ from both sides of the equation to give

$$8y' = x \ .$$

Finally, by dividing by 8, we have isolated the desired variable y':

$$y' = \frac{x}{8} \ . \quad \blacksquare$$

Example 3: Solve for $\dfrac{dy}{dx}$: $x + 2y\dfrac{dy}{dx} = -\dfrac{dy}{dx} + y$.

Solution: Again, isolate the variable we wish to solve for, which in this case is the derivative $\dfrac{dy}{dx}$. Bring all the terms with $\dfrac{dy}{dx}$ to the left, and all others to the right. First, we add $\dfrac{dy}{dx}$ to both sides to give

$$x + 2y\frac{dy}{dx} + \frac{dy}{dx} = y \ .$$

Now subtract x from each side of the equation to give

$$2y\frac{dy}{dx} + \frac{dy}{dx} = y - x \ ,$$

Now we factor out the $\dfrac{dy}{dx}$ to give us

$$\frac{dy}{dx}(2y + 1) = y - x \ .$$

and dividing by $(2y + 1)$ gives the desired result:

$$\frac{dy}{dx} = \frac{y - x}{2y + 1}. \quad \blacksquare$$

Things can get a bit more complicated when the equation contains other variables such as x or y. The method of solution, however, is exactly the same.   Just be very careful when carrying out the steps. Write neatly and slowly.

Example 4:   Solve for $\frac{dy}{dx}$: $5xy + 4\frac{dy}{dx} = 3x^2 - 2xy^2 \frac{dy}{dx}.$

Solution: Bring all the terms with $\frac{dy}{dx}$ to the left by adding $2xy^2 \frac{dy}{dx}$ to both sides of the equation. We get

$$5xy + 4\frac{dy}{dx} + 2xy^2 \frac{dy}{dx} = 3x^2.$$

Now bring all the other terms to the right by subtracting $5xy$ from both sides to get

$$4\frac{dy}{dx} + 2xy^2 \frac{dy}{dx} = 3x^2 - 5xy.$$

On the left, take out the common factor $\frac{dy}{dx}$, to get

$$\frac{dy}{dx}(4 + 2xy^2) = 3x^2 - 5xy.$$

Division by $(4 + 2xy^2)$ gives the result:

$$\frac{dy}{dx} = \frac{3x^2 - 5xy}{4 + 2xy^2}. \quad \blacksquare$$

Remark: Notice that when you add or subtract to move terms to one side or the other, they change sign as they "cross the equal sign." Also notice that when you divide [for example by $4 + 2xy^2$ in the last example], you <u>do not</u> change the sign.

Example 5: Solve for y': $x^2 y' - 2xy + 2xyy' = (x^2 + 1)y'$.

Solution: Again, move all the terms with y' to the left, and all others to the right:

First, subtract $(x^2 + 1)y'$ from both sides, to get

$$x^2 y' - 2x \cdot y + 2xyy' - (x^2 + 1)y' = 0.$$

Then add $2xy$ to both sides of the equation to give

$$x^2 y' + 2xyy' - (x^2 + 1)y' = 2xy.$$

Now, factor out the common factor y':

$$\left(x^2 + 2xy - x^2 - 1\right)y' = 2xy.$$

Notice that the x^2-terms cancel, so divide by what's left, $\left(2xy - 1\right)$, to get

$$y' = \frac{2xy}{\left(2xy - 1\right)}.$$

Note: You can't cancel the $2xy$ term in any way! (If you don't know why, look up "creative canceling" in the index.) ∎

Exercises 7.1

1) Solve for $\dfrac{dy}{dx}$: $4 + 6\dfrac{dy}{dx} = 2 + 4\dfrac{dy}{dx}$.

2) Solve for y': $-y' - 1 = 2y' - 6$.

3) Solve for $\dfrac{dy}{dx}$: $x + y\dfrac{dy}{dx} = 2x + 4y\dfrac{dy}{dx}$.

4) Solve for y': $x^2 - x^2 y' - 1 = xyy' - 6x$.

5) Solve for $\dfrac{dy}{dx}$: $2xy + x^2\dfrac{dy}{dx} + 3x^2y^3 + 3x^3y^2\dfrac{dy}{dx} = 0.$

6) Solve for $\dfrac{dy}{dx}$: $xy + x\dfrac{dy}{dx} + 2xy^2 + 2x^2y\dfrac{dy}{dx} = 3x - 2y\dfrac{dy}{dx}.$

7) Solve for $\dfrac{dy}{dx}$: $y^2\dfrac{dy}{dx} - 2xy = x + 4x^2\dfrac{dy}{dx}.$

Chapter 8

Word Problems

8.1 The Geometry of Rectangles, Triangles, and Circles

Here is a quick review of the connections between various quantities concerning these basic shapes. Consider the rectangle shown.

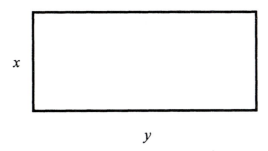

Suppose the width and length of a rectangle are x and y, respectively. Then the area A, the amount of space inside the rectangle, is given by

$$\boxed{A = x\,y}$$

The perimeter P, which is the length of the measurement around the rectangle, is given by

$$\boxed{P = 2x + 2y}$$

For triangles, things get a little more complex. Consider a triangle whose base b is known, and whose height h is also known. What is the area?

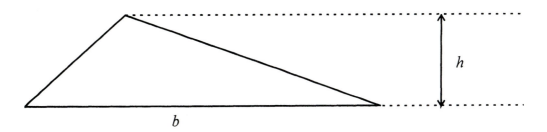

Here is the trick: place the triangle into the following rectangle, and the answer becomes clear.

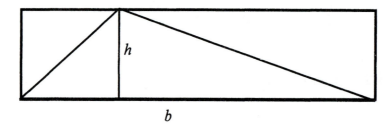

The area of the rectangle is clearly $b \cdot h$, and the area of the triangle is exactly half of the area of the rectangle. (Right? Check the picture.) Hence the area of the triangle is given by

$$A = \frac{1}{2}bh$$

The perimeter of the triangle is tougher, and won't be done here. It is easily done using trigonometry. However, there is one important fact we do wish to mention.

<u>The Pythagorean Theorem:</u>

In $\triangle ABC$,

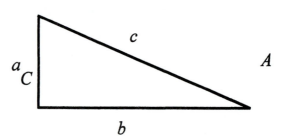

C is a right angle if, and only if,

$$a^2 + b^2 = c^2.$$

(We will not prove this theorem.)

Lastly, consider the circle of radius r.

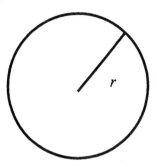

The area of the circle is given by

$$\boxed{A \;=\; \pi\, r^2}$$

which is NOT obvious at all. In the case of circles, the term perimeter is never used. Instead, the word circumference is used, and given by

$$\boxed{C \;=\; 2\,\pi\, r}$$

which is also NOT obvious. It's best to memorize these two equations. Notice that the area involves r^2, while the circumference involves r, which makes sense from the point of view of dimensions: Area is a two-dimensional measure (square feet, square inches, etc.), while circumference is a one-dimensional measure. Knowing that the diameter $D \;=\; 2r$, we can say that $r \;=\; \dfrac{D}{2}$, and so

$$A \;=\; \pi\, r^2 \;=\; \pi \left(\frac{D}{2}\right)^2 \;=\; \frac{\pi}{4} D^2$$

and

$$C \;=\; 2\,\pi\, r \;=\; \pi\, D.$$

We'll use all these facts in the next section, and get to do some exercises there.

8.2 Expressing Quantities in Terms of Other Quantities

Word problems are just exercises posed in words, not symbols, nothing more sinister than that. How do you solve word problems? Well, there is no catalog of word problem "types" for you to memorize. Rather, your principal tool is your own power to think. Make sure, first of all, that you know what the words of the question mean, and exactly what is given to you, and what is asked from you. Understanding exactly what is asked is half the battle.

Next, the question will involve quantities that are related to each other in some way. This relation is given in words, and you must translate it into an expression or an equation – just following what the words tell you. We'll do several examples, starting with easy ones and working up in difficulty.

Example 1:   A rectangular field is to be fenced off next to a straight river, with fencing on three sides, with the river's edge making the fourth side. Exactly 100 feet of fencing is to be used. Express the area of the field as a function of its width.

Solution: First, draw a diagram and label the edges.

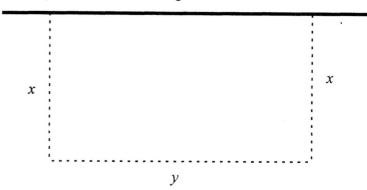

Let x = the width of the field.

Let y = the length of the field.

Let A = the area of the field. We are asked to express A as a function of x. Well, we know $A = xy$, but we need A as a function of x <u>only</u>, not x and y. Notice also that we haven't used the fact that the fencing totals 100 ft, so that

$2x + y = 100$. We can solve this for y in terms of x:

$$y = 100 - 2x.$$

Hence the area $A = xy = x(100 - 2x) = 100x - 2x^2$. ∎

Example 2:   A swimming pool is in the shape of a square with a semicircle at each of two opposite edges. Express the perimeter and area of the pool as a function of the diameter of the semicircles.

Solution:

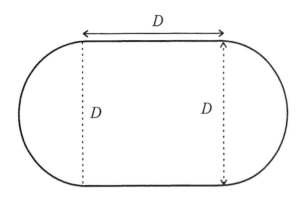

a)   The circumference of a circle is $2\pi r$, so the sum of the arc lengths of the two semicircles is $2\pi r$, which equals πD. The square contributes $2D$, so the perimeter of the pool is $\pi D + 2D$.

b)   The area of a circle is πr^2 , so the sum of the area of the two semicircles is $\pi r^2 = \dfrac{\pi D^2}{4}$, making the area of the pool equal to $\dfrac{\pi D^2}{4} + D^2$.   ■

Example 3:   A cylindrical tin can has height h and radius r. Its volume is 32 cubic centimeters (written cm^3). Express h as a function of r, and vice versa.

Solution: Again, first draw a picture and label the variables.

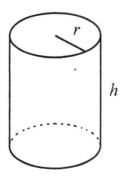

$$V = (\text{ area of end })(\text{height}) = (\pi r^2)h = 32.$$

$$\therefore \ h = \frac{32}{\pi \, r^2} \ , \text{ which is expressed in cm.}$$

$$\text{Also } \ \pi \, r^2 = \frac{32}{h} \ , \text{ so } \ r^2 = \frac{32}{\pi \, h} \ , \text{ and hence}$$

$$r = \sqrt{\frac{32}{\pi \, h}} \ \text{ cm.} \qquad\qquad \blacksquare$$

Example 4:   An equilateral triangle has sides of length s.

a) Express the height as a function of s.

b) Express the area as a function of s.

c) Express the side as a function of area.

Solution:

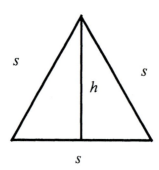

a) We need h as a function of s. It's not obvious how h and s are related, but notice that the altitude is the perpendicular bisector of the base (into two pieces, each of length $\frac{s}{2}$). So consider the left half:

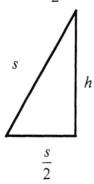

Now we have a right-angled triangle, and can use the Pythagorean theorem.

$$\left(\frac{s}{2}\right)^2 + \ h^2 = s^2$$

Let's simplify and solve for h.

$$\frac{s^2}{4} + h^2 = s^2$$

So
$$h^2 = s^2 - \frac{s^2}{4} = \frac{3}{4}s^2,$$

hence
$$h = \sqrt{\frac{3}{4}}\, s = \frac{\sqrt{3}}{\sqrt{4}}s = \frac{\sqrt{3}}{2}s.$$

b) Well, area $= \dfrac{1}{2}bh$, which in our case gives

$$A = \frac{1}{2}(s)(h)$$

$$= \frac{1}{2}(s)\left(\frac{\sqrt{3}}{2}s\right).$$

So
$$A = \frac{\sqrt{3}}{4}s^2.$$

c) To get s as a function of A, solve this equation for s.

$$s^2 = \frac{4}{\sqrt{3}}A$$

$$s = \pm\sqrt{\frac{4}{\sqrt{3}}A} = \pm\frac{\sqrt{4}}{\sqrt{\sqrt{3}}}\sqrt{A} = \pm\frac{2}{\sqrt[4]{3}}\sqrt{A}$$

But since $s > 0$, we have only $s = \dfrac{2}{\sqrt[4]{3}}\sqrt{A}$. ∎

Example 5:   A 10-inch wire is cut into two pieces. One of the pieces, of length x, is bent into a square. The other piece is bent into a circle. What is the total area of the two shapes as a function of x?

Solution: Since one piece is of length x inches, the other piece is of length $10 - x$ inches.

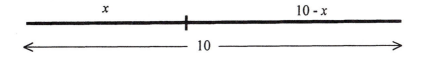

The piece of length x is bent into a square of side $x/4$, and therefore the area of the square is $\dfrac{x^2}{16}$ in². The second piece of wire is of length $10 - x$, which forms the circumference of the circle. Hence we have $10 - x = 2\pi r$, and $r = \dfrac{10 - x}{2\pi}$. So the area of the circle is $\pi r^2 = \pi \left(\dfrac{10 - x}{2\pi}\right)^2 = \dfrac{(10 - x)^2}{4\pi}$.

The sum of the areas is $A = \dfrac{x^2}{16} + \dfrac{(10-x)^2}{4\pi}$ in². ∎

Exercises 8.2

1) A rectangular field of area 20,000 sq ft is to be fenced off next to a river, with fencing on three sides and the river making the fourth side. Express the length of fencing necessary as a function of the width of the field.

2) The shape of a window is given by two squares, one on top of the other, with a semicircle on top of that. Find the perimeter and area of the window as a function of the width of the window.

3) A rectangle is twice as long as it is wide. a) Express the area as a function of the width. b) Express the perimeter as a function of the width. c) Express the area as a function of the perimeter. d) Express the perimeter as a function of the area.

4) A cylindrical can is to be made up from sheet steel so that its surface area is 100 sq in. Express the height as a function of the radius. Hint: Imagine removing the top and bottom with a can opener, and splitting the rest down the side and unrolling it flat.

5) An open cardboard box is to be constructed from a rectangular 8" by 10" sheet by cutting identical squares of side x out of each corner, and folding up the resulting edges. Determine the volume of the box as a function of x.

6) An open cardboard box is to be constructed from a rectangular 1.5 m by 2 m sheet by cutting identical squares of side x out of each corner, and folding up the resulting edges. Determine the volume of the box as a function of x. Also, what is the exterior surface area as a function of x?

7) The volume of a sphere is $V = \dfrac{4}{3}\pi r^3$, and its surface area is $S = 4\pi r^2$. Express V as a function of S and vice versa.

8) A 2 ft wire is cut into two pieces. One of the pieces, of length x, is bent into a circle. The other piece is bent into a rectangle whose length is twice the size of its width. What is the total area of the two shapes as a function of x?

9) Suppose you have a rectangle and you cut off a square from one end of it as shown.

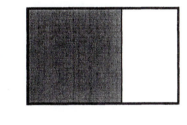

w

l

If the remaining rectangle on the right has the same ratio of $\dfrac{\text{length}}{\text{width}}$, then the rectangle is called a golden rectangle, and that ratio is called the golden ratio.   Find the numerical value of the golden ratio.

10) Sardinia Airlines is ordering some new planes, and deciding how many seats to have in them. The plane can fit 50 seats very comfortably, but for each seat over 50, the airline finds that it must lower the price P of all the tickets by 1% because the passengers are feeling "packed in." a) What is the total maximum revenue per flight, in terms of P, if there are 50 seats? b) What if there are 51 seats? c) 80 seats? d) What if there are x seats, where $x > 50$?

Chapter 9

Exponential and Logarithmic Functions

9.1 Introduction

An exponential function has the form $f(x) = a^x$, where $a > 0$. The number a is called the <u>base</u>. Consider $a = 2$. It is clear what $f(x) = 2^x$ means for some values of x. For example

$$f(0) = 2^0 = 1, \qquad f(1) = 2^1 = 2,$$

$$f(2) = 2^2 = 4, \qquad f(3) = 2^3 = 8,$$

$$f(-1) = 2^{-1} = \frac{1}{2}, \qquad f(-2) = 2^{-2} = \frac{1}{4},$$

$$f(\tfrac{1}{2}) = 2^{\frac{1}{2}} = \sqrt{2} \cong 1.414,$$

and $\qquad f(3.2) = 2^{3.2} = 2^3 \cdot 2^{\frac{1}{5}} = 8\sqrt[5]{2}.$

This last one could be tough to calculate without a calculator, but at least you know <u>what it means</u> (see Section 1.5). Use the above values to plot these points for the graph of $y(x) = 2^x$.

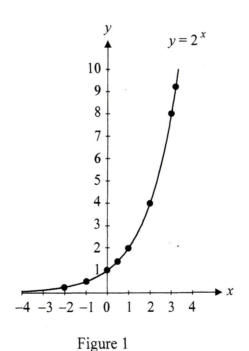

Figure 1

We know what all rational exponents of 2 mean: $2^{m/n} = (\sqrt[n]{2})^m$, if m/n is in lowest terms. What happens at all the remaining (irrational) points? It can be shown (but not here) that there is exactly <u>one</u>

smooth curve, always increasing, that can be drawn through all these points. That is the graph of 2^x, for x real. Notice that its domain is $(-\infty,\infty)$ and the range is $(0,\infty)$.

Graphs of a^x, for $a > 1$

Using these methods, plot the family of graphs a^x. Figure 2 shows several exponential functions for the case when $a > 1$.

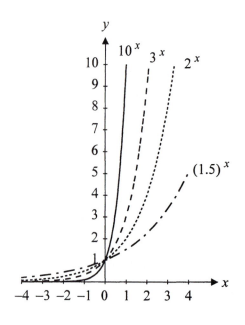

Figure 2

Note: a) As x gets large, each of these functions increases without bound (goes to ∞), but 10^x does it much faster than 3^x, etc.

b) As x goes to $-\infty$, these functions go to 0. But again, 10^x does it much faster than 3^x, etc.

c) All exponential functions pass through the point $(0,1)$.

Graphs of a^x, for $0 < a < 1$

Figure 3 shows several exponential functions for the second case, where $0 < a < 1$, which can be obtained by plotting a few points. (A few thousand that is, if you're the computer plotting this graph.)

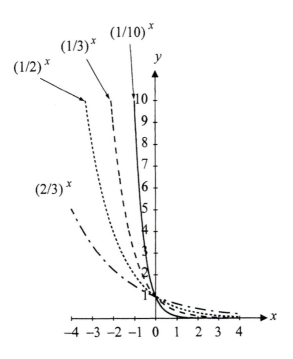

Figure 3

By comparing Figures 2 and 3, it certainly <u>looks</u> like $\left(\dfrac{1}{10}\right)^x$ is the mirror image of 10^x, and $\left(\dfrac{1}{2}\right)^x$ is the mirror image of 2^x, etc. Notice that a function $g(x)$ is the mirror image about the y-axis of $f(x)$ if, and only if, $g(-x) = f(x)$, as shown in Figure 4.

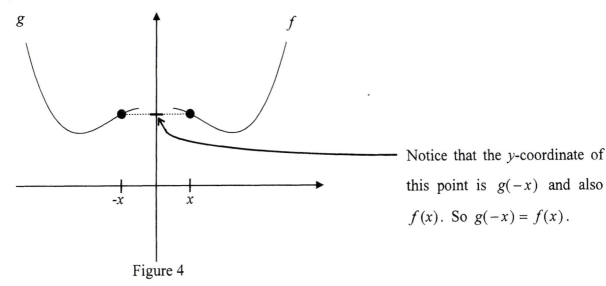

Notice that the y-coordinate of this point is $g(-x)$ and also $f(x)$. So $g(-x) = f(x)$.

Figure 4

Let $f(x) = 2^x$ and $g(x) = \left(\dfrac{1}{2}\right)^x$. Then $g(-x) = \left(\dfrac{1}{2}\right)^{-x}$. Recalling the exponent laws from Chapter

1, $\left(\dfrac{1}{2}\right)^{-x} = \left(\dfrac{2}{1}\right)^x = 2^x = f(x)$, so that $g(-x) = f(x)$. Hence $\left(\dfrac{1}{2}\right)^x$ is the reflection about the y-axis of

2^x. This is shown in Figure 5.

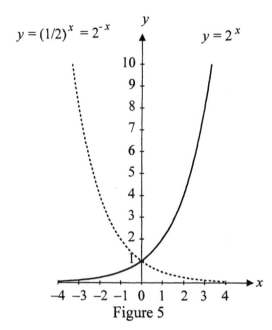

Figure 5

By the way, notice that a^x is defined <u>for all x</u> only if $a > 0$. If $a < 0$, you can no longer have a^x for all x. For example if $a = -1$ and $x = \frac{1}{2}$, we have $(-1)^{\frac{1}{2}} = \sqrt{-1}$, which is not a real number. So the family of functions a^x is defined for $a > 0$ and any real number x.

Exercises 9.1

1) Determine the behavior of the following exponential functions as $x \to \pm\infty$, then sketch the graph of the function, labeling at least three points.

     a) $f(x) = \left(\dfrac{2}{3}\right)^x$ b) $f(x) = \left(\dfrac{3}{2}\right)^x$

     c) $f(x) = 1.1^x$ d) $f(x) = .32^x$

2) Graph the function $f(x) = 2^x$, then using that result and the methods learned in Chapter 3, graph the following functions:

     a) $f(x) = 2^{x-1}$ b) $f(x) = 2^{x+3}$ c) $g(x) = -2^{x+2}$ d) $f(x) = -2^{x-1}$

3) Graph the function $f(x) = 3^x$, then using that result, graph the following functions:

     a) $f(x) = 3^{x+1}$ b) $f(x) = -\frac{1}{2} \cdot 3^{x+1}$

9.2 <u>The Function e^{x} – also called THE Exponential</u>

Of all the exponential functions a^x, the one that has a $45°$ tangent at $x = 0$ is especially important. This exponential is depicted in Figure 6.

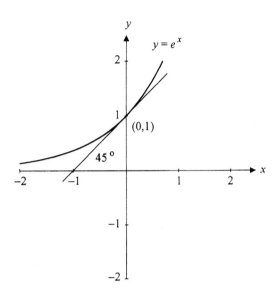

Figure 6

You can tell from the graphs of 2^x and 3^x that this special function lies between them.   The particular value of a that gives this exponential is called "e." It can be calculated that $e \cong 2.718$. (In fact, $e \cong 2.718281828459045\ldots$.) The exact reason why e^x is so important becomes clear in calculus, when you see that of all a^x, e^x has the simplest derivative.

Since any function a^x can be written as e^{kx}, for some k, that is the form ordinarily used in technical work. (See Section 9.9 for details.)

Exercises 9.2

1) Sketch: a) $y = e^{x-1}$ b) $y = e^{-x}$ c) $y = -\frac{1}{2}e^x$

               d) $f(x) = e^x - 1$ e) $y = -\frac{1}{2}e^{-x}$

2) Sketch e^{-x^2}. (<u>Hint:</u> Plot and think!)

9.3 The Idea of Inverses

Before discussing inverses, we remark that there are many different ways of talking about functions and their operations. For example, consider the function $f(x) = x^3$ at the point $x = 2$. Then all of the following ways of speaking mean the same thing.

1. Taking f of 2 gives you 8.

2. Evaluating f at 2 yields 8.

3. f operating on 2 gives 8.

4. Applying f to 2 gives 8.

5. f takes 2 to 8.

6. f acting on 2 yields 8.

7. $f(2) = 8$.

8. $2 \overset{f}{\longmapsto} 8$

9. $f : 2 \longmapsto 8$

10. (2,8) is a point on the graph of f.

All of these ways of saying the same thing are actually used. They all have their advantages in various different contexts.

When one function undoes the action of another, it is said to be the _inverse of the other_. For example, look at $f(x) = x^3$ and $g(x) = \sqrt[3]{x}$. If you take any number, x, cube it, and then take the cube root of the result, you're back to x. (Try this for $x = 2$ and 3.) In symbols

$$\sqrt[3]{x^3} = x \quad \text{or} \quad g(f(x)) = x.$$

Similarly, you can show $\qquad \left(\sqrt[3]{x}\right)^3 = x \quad \text{or} \quad f(g(x)) = x.$

A second example is the doubling function, $f(x) = 2x$, and the halving function, $g(x) = \dfrac{x}{2}$; they're inverse to each other. The function $f(x) = \dfrac{1}{x}$ is its own inverse!

Definition: a) We say $f(x)$ and $g(x)$ are _inverse to each other_ if $f(g(x)) = x$ and $g(f(x)) = x$, and if domain of f = range of g and domain of g = range of f.

b) The inverse of a function f is denoted f^{-1}.

Note that the domains and ranges are important to the discussion of inverse functions. Two function expressions can be inverses over one interval but not inverses over another interval. (See Exercise 8 in

Section 9.5.) The notation for the inverse, f^{-1}, is standard, but it is an unfortunate choice because it <u>looks</u> like f to the power -1. So, if you really want to say f to the power -1, you should write $\dfrac{1}{f}$.

Inverses, and finding them, are a big deal in mathematics. Here's just a little example. Suppose you wish to solve the equation $f(x) = 0$. If you could find f^{-1}, you could apply it to both sides to get $f^{-1}(f(x)) = f^{-1}(0)$, and so $x = f^{-1}(0)$. Notice that the left side reduces to x, so PRESTO! You've solved the equation.

The question is: Okay, suppose that you've got a function f, how do you find f^{-1}? Specifically: if you've got the graph of f, how do you find the graph of f^{-1}? Or, if you've got an expression for f, how do you get the expression for f^{-1}? Read on!

9.4 Finding the Inverse of f Given by a Graph

First of all, notice that the graph of f is the set of points of the form $(x, f(x))$. (Do you agree?) For every x in the domain of f, f takes x to $f(x)$. Use the following symbols:

$$x \quad \overset{f}{\longmapsto} \quad f(x)$$

Notice that f^{-1} takes $f(x)$ to x. So

$$f(x) \quad \overset{f^{-1}}{\longmapsto} \quad x$$

Hence the graph of f^{-1} is exactly the set of points $(f(x), x)$. (Notice, for example, that the point $(2,8)$ is on the graph of x^3, and $(8,2)$ is on the graph of $\sqrt[3]{x}$.) The result is that the graph of f^{-1} is just the graph of f with the order of the coordinates reversed. How do you do that, you ask? The picture below tells all!

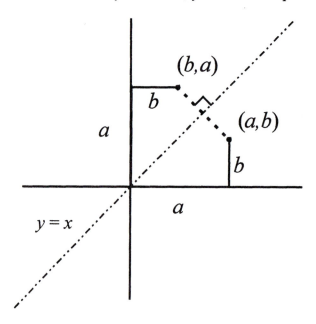

Figure 7

Using geometry, you can prove that (a,b) and (b,a) are reflections of each other about the line $y = x$, that is, the line through the origin at an angle of $45°$. So to go from the graph of f to the graph of f^{-1}, simply reflect the entire graph of f about the line $y = x$.

Figure 8 shows the inverse functions $f(x) = 2x$ and $g(x) = \dfrac{x}{2}$.

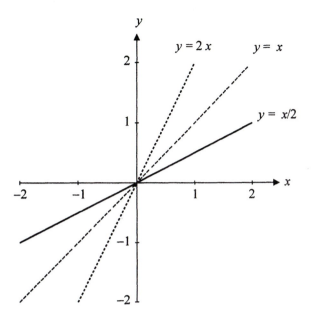

Figure 8

Notice that each graph is obtained from the other by reflecting across the line $y = x$.

Consider the functions $f(x) = x^3$ and $g(x) = \sqrt[3]{x}$. They are also reflections of each other about the line $y = x$.

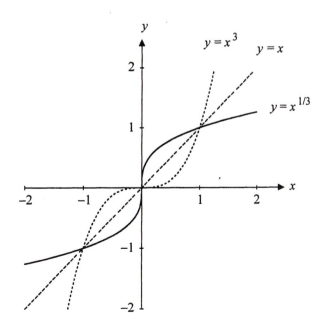

Figure 9

Example 1:   Show that the functions $f(x) = x+2$ and $g(x) = x-2$ are inverses of one another on the interval $(-\infty, \infty)$, and then graph the functions.

Solution: Since $f\big(g(x)\big) = f(x-2) = (x-2)+2 = x$ and

$g\big(f(x)\big) = g(x+2) = (x+2)-2 = x$ for any value of x, these two functions are inverses on the entire interval $(-\infty, \infty)$. The graphs are shown below.

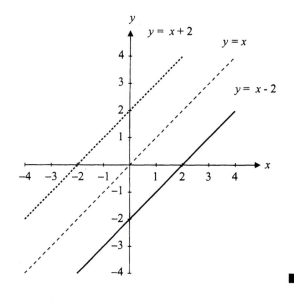

The function $f(x) = \dfrac{1}{x}$ is its own inverse since it reflects back onto itself, as shown in Figure 10.

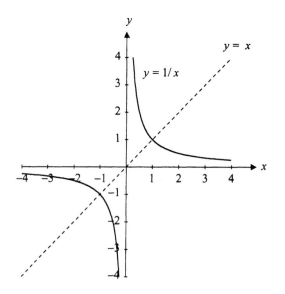

Figure 10

Notice that some functions don't have inverse functions. Here's an example.

Example 2:   Show graphically that the function $f(x) = x^2$ does not have an inverse function on the interval $(-\infty, \infty)$.

Solution: A picture's worth a thousand words.

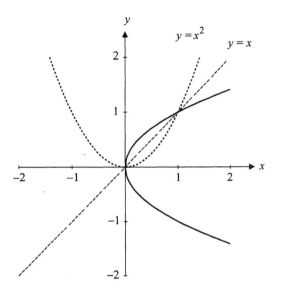

You can flip the graph of $f(x) = x^2$ but the resulting graph (given by the solid curve) is NOT a function. Think about it: how would you evaluate it at $x = 1$, for example? There are two candidates, 1 and −1. A function can have at most one value for each x. In other words, each vertical line can cross the graph of a function at most once. That's called the "<u>vertical line test for being a function</u>." We conclude that $f(x) = x^2$ on $(-\infty, \infty)$ does not have an inverse function. ■

We got into this mess because our original function $f(x) = x^2$ had instances where two different x-values had the same $f(x)$-value. (What's an easy example of this?) That is, its graph was crossed by horizontal lines more than once each. It flunked the "<u>horizontal line test for a function to have an inverse</u>." Any function whose graph is crossed by a given horizontal line at most once is called <u>invertible</u> – that is, f^{-1} exists. Another term for invertible is <u>one-to-one</u>, written 1-1. (Why is this a good name?) So the problem was that $f(x) = x^2$ is not 1-1. But suppose you really wanted an inverse for $f(x) = x^2$. You could (a) get depressed or (b) settle for a piece of the whole thing, as in the next example.

Example 3: Consider the function $f(x) = x^2$, for $x \geq 0$. (Notice the restricted domain.) Graph its inverse.

Solution: No problem!

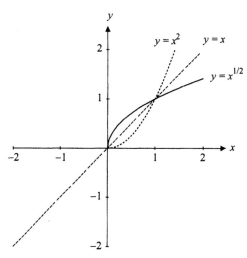

All the tests are passed. That inverse, by the way, is $g(x) = x^{1/2} = \sqrt{x}$. ■

Exercises 9.4 In Exercises 1–8, graph the given functions, and their inverses, if they exist. If they do not, explain why.

1) $f(x) = 2x - 1$ 2) $f(x) = x^5$ 3) $g(x) = -x$

4) $k(y) = y^4 + 1$ 5) $h(x) = 2^x$ 6) $L(s) = 1^s$

7) $f(x) = \left(\dfrac{1}{2}\right)^x$ 8) $f(x) = x^3 - 1$

9) Is the function $f(x) = 1 - x^2$ invertible for $-1 \leq x \leq 1$? How about for the interval $0 \leq x \leq 1$? (Sketch!)

9.5 Finding the Inverse of *f* Given by an Expression

Suppose we have $f(x)$ given by an expression, in other words we have the equation $y = f(x)$, and suppose we want an expression for $f^{-1}(x)$. If we know f is invertible then f^{-1} exists, even though we don't yet know what it looks like. In any event, if we apply f^{-1} to both sides, we get

$$f^{-1}(y) = f^{-1}\big(f(x)\big),$$

and so
$$f^{-1}(y) = x.$$

But if you solved the equation $y = f(x)$ for x, you would get $x =$ some function of y. Hence $f^{-1}(y)$ is that function, meaning that to find $f^{-1}(y)$ you just solve the equation $y = f(x)$ for x. So here's the complete method.

1) Write down the equation $y = f(x)$.

2) Solve for x. (This gives you $x = f^{-1}(y)$.)

3) Take the expression $f^{-1}(y)$. Pluck out the y, put in the x, and Voila! It's ready to eat.

Example 1:   Find the inverse of the function $f(x) = \dfrac{2}{x+3}$.

Solution: Write $y = \dfrac{2}{x+3}$ and solve this equation for x to obtain $f^{-1}(y)$. Multiplying by $x + 3$ gives

$$xy + 3y = 2,$$

which in turn means
$$xy = 2 - 3y,$$

and hence
$$x = \frac{2 - 3y}{y} = f^{-1}(y).$$

Using x instead of y gives

$$f^{-1}(x) = \frac{2 - 3x}{x}. \blacksquare$$

You can check this result by determining whether $f\!\left(f^{-1}(x)\right) = x$ and $f^{-1}\!\left(f(x)\right) = x$. In this case

$$f^{-1}\!\left(f(x)\right) = f^{-1}\!\left(\frac{2}{x+3}\right)$$

$$= \frac{2 - 3\left(\dfrac{2}{x+3}\right)}{\dfrac{2}{x+3}} = \frac{\dfrac{2x+6-6}{x+3}}{\dfrac{2}{x+3}} = \frac{\dfrac{2x}{x+3}}{\dfrac{2}{x+3}} = \frac{2x}{2} = x.$$

Also

$$f\!\left(f^{-1}(x)\right) = f\!\left(\frac{2-3x}{x}\right)$$

$$= \frac{2}{\dfrac{2-3x}{x}+3} = \frac{2}{\dfrac{2-3x+3x}{x}} = \frac{2}{\dfrac{2}{x}} = x.$$

Check! We're done.

Example 2:   Find the inverse of $f(x) = \sqrt[3]{2x+1}$.

Solution: Write $y = \sqrt[3]{2x+1}$, and solve for x.

$$y^3 = 2x + 1$$

$$2x = y^3 - 1$$

$$x = \frac{y^3 - 1}{2} = f^{-1}(y),$$

$$\therefore f^{-1}(x) = \frac{x^3 - 1}{2} \quad \blacksquare$$

Question: It all looks so easy. Can anything go wrong? And what if a given function doesn't <u>have</u> an inverse? How will that show up?

Answer: Yes, things can deteriorate. For example, what if you can't solve for x? If that happens, then you're stuck. Moreover, if f^{-1} doesn't exist, it will show up by the fact that the equation <u>cannot</u> be solved for x. Not even by Gauss, with help from Einstein. See the next example.

Example 3:   Find the inverse of $f(x) = x^4 - 3$.

Solution: Write $y = x^4 - 3$, and solve for x.

$$x^4 = y + 3$$

$$x = \pm \sqrt[4]{y+3}$$

Aha, you see, that's not a function. You have not solved for x as a function of y. Of course, you knew that $f(x)$ does not have an inverse, because it flunks the horizontal line test. ∎

Exercises 9.5   In Exercises 1-6, find inverses, if they exist, of the given functions. If they do not, explain why.

1) $f(x) = 2x - 3$ 　　2) $k(x) = \dfrac{x}{x+1}$ 　　3) $g(x) = \sqrt[3]{5x + 1}$

4) $s(t) = \sqrt{t + 2}$ 　　5) $f(x) = \dfrac{2}{x}$ 　　6) $f(w) = \dfrac{w^2}{w^2 + 1}$

7)   The function $f(x) = (x-1)^4$ does not have an inverse on the interval $(-\infty, \infty)$. Show this. Then show that, if you restrict the domain to $[1,\infty)$, this restricted function has an inverse. Graph both functions.

8)   a) Let $f(x) = x^2 + 3$ on $[0,1]$. Find its domain and range and then sketch it and its inverse. Find an expression for the inverse $f^{-1}(x)$.

b) Let $g(x) = x^2 + 3$ on $[-1,0]$. Do the same as in part (a).

c) The function expressions of f and g are the same. Are their inverses the same?

9.6 <u>Definition of Logarithms</u>

There's one more function you'll meet in science, engineering, management, the life sciences, and sometimes in the newspaper. In symbols, it looks like this:

$$\log_a x.$$

It is read as "<u>log, to the base *a*, of *x*.</u>" (Do NOT read this as "the log of *a*-to-the-*x*!") There are several ways of defining it. Here's one.

<u>Definition:</u>   Let $a > 0$, $a \neq 1$. Then $\log_a x$ is <u>the number to which you raise *a* to get *x*.</u>

Example 1:   Show that $\log_2 8 = 3$.

   Solution: Here the base is 2 and $x = 8$. To what number do you have to raise 2 in order to get 8? Answer: 3, so $\log_2 8 = 3$. ■

Example 2:   Show that $\log_{10} 1,000,000 = 6$.

   Solution: Here the base is 10, and $x = 1,000,000$. What number do you have to raise 10 to, in order to get 1,000,000 (6 zeros)? Answer: 6, so $\log_{10} 1,000,000 = 6$. ■

Example 3: $\log_{10} .01 = ?$

   Solution: Here the base is 10, and $x = .01$. Write .01 as a power of 10. Here you go:

$$.01 = \frac{1}{100} = \frac{1}{10^2} = 10^{-2}$$

What number do you have to raise 10 to in order to get 10^{-2}? Answer: -2, of course, so $\log_{10} .01 = -2$. ■

Example 4: $\log_2 32 = ?$

Solution: Write 32 as a power of 2:

$$32 = 2 \cdot 2 \cdot 2 \cdot 2 \cdot 2 = 2^5,$$

so, $\log_2 32 = 5.$ ■

Example 5: $\log_3 81 = ?$

Solution: Write 81 as a power of 3:

$$81 = 3 \cdot 3 \cdot 3 \cdot 3 = 3^4,$$

so, $\log_3 81 = 4.$ ■

Example 6: $\log_7 7^{15} = ?$

Solution: 7^{15} is already written as a power of 7. What a silly question! $\log_7 7^{15} = 15.$ ■

Example 7: $\log_a a^3 = ?$

Solution: Another silly question! If you write a^3 as a power of a, obviously the exponent must be 3, so $\log_a a^3 = 3.$ ■

Remark: The function $\log_{10} x$ is known as the <u>common logarithm</u>. Sometimes you will see the expression $\log x$. Usually that means $\log_{10} x$, although in some advanced mathematics topics it stands for $\log_e x$.

Exercises 9.6

Evaluate the following:

1) $\log_9 81$

2) $\log_2 \dfrac{1}{2}$

3) $\log \dfrac{1}{1000}$

4) $\log_{\frac{1}{2}} 8$

5) $\log_3 \sqrt{3}$

6) $\log_4 2\sqrt{2}$

7) $\log_b b^{13}$

8) $\log_a a^x$

9.7 Logs as Inverses of Exponential Functions; Graphs and Equations

You will see that there is a fundamental relationship between logs and exponentials: they are inverse to each other. Recall from Section 9.3 that f and g are called inverse to each other if all the following are true:

a) $f(g(x)) = x$,

b) $g(f(x)) = x$,

c) domain of f = range of g,

and

d) domain of g = range of f.

Theorem: Let $a > 0$, $a \neq 1$. Then $\log_a x$ and a^x are inverse to each other.

Remark: This theorem has been proved rigorously using the continuity properties from calculus, but that cannot be done here. However, if you let $f(x) = a^x$ and $g(x) = \log_a x$, you can verify the first two conditions.

a) $f(g(x)) = a^{g(x)} = a^{\log_a x}$. What does this mean? It is a, raised to the number to which you raise a to get x. So it equals x – that is, $f(g(x)) = a^{g(x)} = a^{\log_a x} = x$.

b) $g(f(x)) = \log_a f(x) = \log_a a^x = x$. (Just like Examples 6 and 7 in the previous section!)

So you haven't completely proved this theorem, but you can see that $\log_a x$ and a^x undo each other.

Knowing that $\log_a x$ and a^x are inverses allows you immediately to graph $\log_a x$. If you wish to graph the function $f(x) = \log_2 x$, you need only graph the function $g(x) = 2^x$, and flip it around the line $y = x$ (see Figure 11).

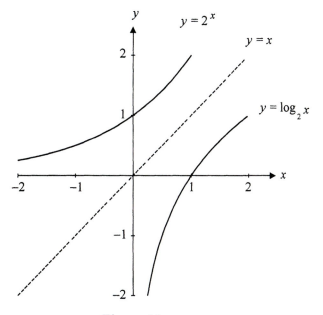

Figure 11

Notice that the domain of $f(x) = \log_2 x$ is the set of all positive numbers, and the range is the set of all numbers. Notice also $\log_2 1 = 0$. Figure 12 shows the common logarithm.

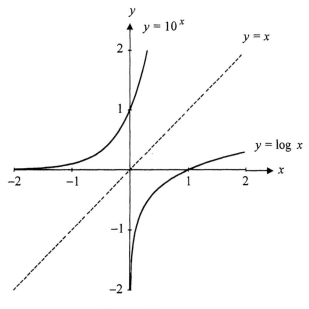

Figure 12

As with other functions, once you know what the log graph looks like, you can obtain other graphs by shifting. Figure 13 shows how the graph of $\log_2 (x+3)$ is obtained from $\log_2 x$ by shifting it to the left 3 units.

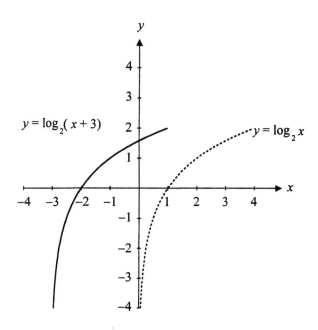

Figure 13

By recalling the definition of logs and their relationship to exponentials, it is possible to solve some new equations containing such curiosities. The following examples show some of the problems you may encounter.

Example 1:   Solve $\log_2 x = 4$.

Solution: We know that the function 2^x <u>undoes</u> the action of that log function, and so if we apply it to the left side of the equation we get x. This means that if we apply 2^x to both sides of the equation, we get

$$x = 2^4 = 16.$$

(The left side is x because the log was "undone" by the action of 2^x; this is the whole point of "inverseness," which is how $\log_2 x$ and 2^x are related.) ∎

Example 2:   Solve $\log_{10} x = 3$.

Solution: As in the last example, to "undo" the action of the log (now with base 10) apply 10^x to both sides. (Recall that applying 10^x to 3 means evaluating 10^x at 3.)

So $x = 10^3 = 1000$. ∎

Example 3:   Solve $\log_{10}(x^2 - 4x + 14) = 1$.

Solution: Again, apply 10^x to both sides. So,

$$(x^2 - 4x + 14) = 10^1 = 10$$

or $x^2 - 4x + 4 = 0,$

which factors as $(x - 2)^2 = 0$. So $x = 2$. ■

Example 4:   Solve $\log_3(x^2 - 3x - 7) = 1$.

Solution: To "peel off" the log, apply 3^x, and in doing so you get

$$(x^2 - 3x - 7) = 3^1 = 3$$

or $x^2 - 3x - 10 = 0.$

Factor this last equation to give

$$(x - 5)(x + 2) = 0,$$

which has solutions $x = 5$ and $x = -2$. ■

Example 5:   Solve $2^{x^2 + 1} = 8$.

Solution: Apply $\log_2 x$ to both sides (that is, take $\log_2$ of both sides) and get

$$x^2 + 1 = \log_2 8$$

or $x^2 + 1 = 3.$

This gives

$$x^2 = 2$$

or $x = \pm\sqrt{2}$. ■

Exercises 9.7

1) a) Graph $f(x) = 2^{x+1}$.

   b) Show that $g(x) = \log_2 \dfrac{x}{2}$ is the inverse of $f(x)$ and graph it.

2) Using a calculator, graph $y = \log x$ between .5 and 2 by using several x-values. Using the same x-values, graph $y = \log(10x)$. What do you notice?

3) Solve the following:

   a) $2^{x-3} = 64$ b) $3^{x+1} = 27$ c) $4^{2x-3} = 16$ d) $5^{x+5} = \dfrac{1}{125}$

4) Solve $\log_3 (x + 7) = -1$.

5) Solve $\log_{64} x^2 = \dfrac{1}{3}$.

6) Solve $\log_3 (x^2 - 5) = 2$.

9.8 __Laws of Logarithms__

Your life among the logs is made much simpler when you know certain log laws. They're great when you're solving equations or simplifying expressions.

__Log Laws:__   Let $a > 0$, $a \neq 1$, and $x > 0$ and $y > 0$. Then

1) $\log_a xy = \log_a x + \log_a y$

2) $\log_a \dfrac{x}{y} = \log_a x - \log_a y$

3) $\log_a x^r = r \log_a x$

4) $\log_a 1 = 0$ for all a (but we knew that already!)

5) For any $a > 0$, with $a \neq 1$, $\log_a x = \dfrac{\log_b x}{\log_b a}$, for any convenient b – for example, 10 or e.

This "change of base" law can be a lifesaver if you can't handle $\log_a x$.

Example 1:   Solve $\log_2 x^2 + \log_2 2x = 4$.

Solution: We can combine the left side using rule 1.

So $\log_2 2x^3 = 4$,

using the first law.

Now we can apply 2^x to undo the log (before we couldn't):

$$2x^3 = 2^4 = 16$$

$$x^3 = 8$$

$$x = 2.$$

We should check this solution:
$$\log_2 (2^2) + \log_2 (2\cdot 2) = \log_2 4 + \log_2 4 = 2 + 2 = 4. \quad \blacksquare$$

Example 2:   Solve $\log_{10} (x^2 - 3x)^3 = 3$.

Solution: Using the third law, you obtain

$$3\log_{10} (x^2 - 3x) = 3, \text{ or}$$

$$\log_{10} (x^2 - 3x) = 1. \text{ Now apply } 10^x:$$

$$x^2 - 3x = 10^1 = 10.$$

Hence $x^2 - 3x - 10 = 0,$

$$(x-5)(x+2) = 0,$$

and $x = 5 \text{ or } -2.$ ∎

In this example, you calculated two answers, both of which were valid because they satisfied the original equation. Sometimes, however, you may get some answers that are not valid. (They are called extraneous solutions.) You must check your solutions to determine that they are valid. Consider the following.

Example 3:   Solve $\log_2 x + \log_2 (x-1) = 1$.

Solution: Using the first law, we can combine those two logs to get

$$\log_2 \big(x(x-1)\big) = 1.$$

Now apply 2^x and get

$$x(x-1) = 2^1 = 2.$$

Hence $x^2 - x - 2 = 0$

or $(x-2)(x+1) = 0,$

giving $x = 2 \text{ or } -1.$

However, when checking these "solutions," notice that the initial equation is not satisfied for $x = -1$, so $x = 2$ is the only solution. ∎

Example 4:   Evaluate $\log_7 5$.

Solution: We can't evaluate this directly. We could use a calculator, but it doesn't have $\log_7 x$. But, by the change of base law, we get

$$\log_7 5 = \frac{\log_{10} 5}{\log_{10} 7},$$

so $$\log_7 5 \cong \frac{.699}{.845} \cong .827. \blacksquare$$

Exercises 9.8

1) Solve: $\log_3 x + \log_3 (x-6) = 3$.

2) Solve: $\log y + \log y^2 = -1$.

3) Solve: $\log_6 (2x + 1) - \log_6 (2x - 1) = 1$.

4) Solve: $\log_2 x^2 - \log_2 (3x - 8) = 2$.

5) Find numbers a, x, and y where the value of $\log_a (x + y)$ is not equal to the value of $\log_a x + \log_a y$. (<u>Hint:</u> There are many such numbers.)

6) Approximate $\log_3 4$ using the change of base formula and your calculator.

9.9 <u>The Natural Logarithm</u>

In Section 9.2, we introduced the exponential function e^x. Its inverse, $\log_e x$, is called the <u>natural logarithm</u>. For simplicity, $\log_e x$ is denoted by the symbol $\ln x$. To get its graph, we flip e^x about the line $y = x$.

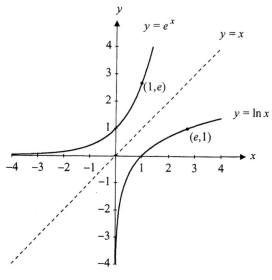

Figure 14

Notice that $\ln 1 = 0$ and $\ln e = 1$. You will also encounter the natural logarithm when you study integration. It helps to know about $\ln x$ and how to graph related functions. Figure 15 shows several horizontal shifts of the natural logarithm.

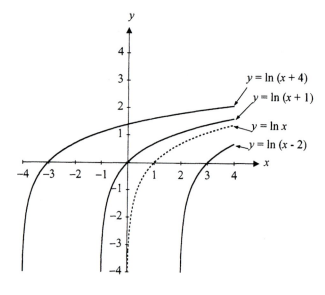

Figure 15

In solving equations, you handle the natural log the same as any other log, keeping in mind that the base is that special number e.

Example 1:   Solve $\ln(x^2 - 1)^3 = 1$.

Solution: Apply e^x to get

$$(x^2 - 1)^3 = e$$

or $$(x^2 - 1) = e^{\frac{1}{3}},$$

which can be solved to give

$$x = \pm\sqrt{1 + e^{\frac{1}{3}}} \,.$$

For most purposes you can leave your answer in this form. If its decimal approximation is needed, you can use your calculator to determine one. ∎

Remark: As mentioned in Section 9.2, any exponential a^x can be written in the form e^{kx} for some constant k. The TRICK: write a as $e^{\ln a}$. So

$$a^x = (e^{\ln a})^x = e^{(\ln a)x} \quad (\text{Recall: } (a^m)^n = a^{mn}).$$

That's it! Now consider the following:

Example 2:   Write 2^x in the form e^{kx}.

Solution: $2^x = (e^{\ln 2})^x = e^{(\ln 2)x}$,

and since $\ln 2 \cong .693$,

$$2^x \cong e^{.693x}. \quad ∎$$

Example 3:   Write $(.345)^x$ in the form e^{kx}.

Solution: $(.345)^x = (e^{\ln .345})^x = e^{(\ln .345)x}$,

and since $\ln .345 \cong -1.06$,

$$(.345)^x \cong e^{-1.06x}. \quad ∎$$

<u>Remarks:</u> a) If you convert the function a^x into the form e^{kx}, then if $a > 1$, as in Example 2, the constant $k > 0$. All functions of the form e^{kx}, for $k > 0$, look similar to the graph shown in Figure 16.

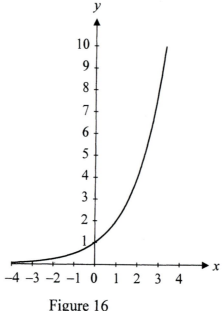

Figure 16

In this case, the function is said to have <u>exponential growth</u>. An example from biology would be the growth of a colony of bacteria under ideal conditions.

b) On the other hand, if $0 < a < 1$, as in Example 3, then the constant $k < 0$. All functions of the form e^{kx}, for $k < 0$, look similar to the graph shown in Figure 17.

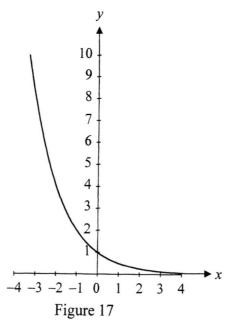

Figure 17

In this case, the function is said to have <u>exponential decay</u>. In physics, the decay of radioactive substances is represented by such functions.

Exercises 9.9

1) Graph e^{x-1}.

2) Graph $y = \ln(x-1)$.

3) Graph $y = 2 + \ln(x+1)$.

4) Solve $\ln t + \ln t^2 = 6$.

5) Solve $e^{x^2 + 2x - 3} = 1$.

6) Solve $e^{\ln(w^2 + 1)} = 5$. ($\leftarrow$ Gift!)

7) Write 10^x and $(0.5)^x$ in the form e^{kx}. (Use a calculator.)

Appendix A

Completing the Square and Deriving the Quadratic Formula

A.1 Completing the Square

A quadratic expression in x is of the form $ax^2 + bx + c$. It's a polynomial of degree 2. One important way of changing the form of a quadratic is called completing the square. It is one of the most frequently used methods of changing the form of an expression to suit a particular purpose. It is used for graphing circles, ellipses, parabolas, and hyperbolas, for deriving the quadratic formula and integrating certain functions, and for many other purposes. You'll meet them soon enough. We'll do an example first, and then look at the general method.

Example 1: Complete the square for $f(x) = x^2 + 8x + 12$.

Solution: (First, notice that the x^2 coefficient is 1. If it is not, it must be factored out.) We take 8, the coefficient of x, take half of the 8 to get 4, square the 4 to get 16, which we add (and subtract), to get

$$f(x) = (x^2 + 8x + 16) + (12 - 16).$$

The first term is a perfect square. (That was the whole idea – and that's why it's called "completing the square.") So

$$f(x) = (x+4)^2 - 4. \blacksquare$$

Here's the general method for completing the square of $f(x) = ax^2 + bx + c$:

a)   If $a \neq 1$, factor out the a from the first two terms to get $f(x) = a\left(x^2 + \dfrac{b}{a}x\right) + c$.

b)   Take half of the coefficient of the resulting x-term, and square it.

c)   Add and subtract that number (inside the parentheses if $a \neq 1$).

d)   Rewrite $f(x)$ as the sum or difference of a perfect square and a number.

Example 2:   Complete the square for $f(x) = x^2 - 3x + 4$.

Solution: Here, the coefficient of the x^2-term is 1, which means we can skip the first step. The x-coefficient is -3, half of it is $-\dfrac{3}{2}$ and squaring it gives $\dfrac{9}{4}$. Adding and subtracting $\dfrac{9}{4}$ gives us

$$f(x) = \left(x^2 - 3x + \frac{9}{4}\right) + \left(4 - \frac{9}{4}\right).$$

The first term is the perfect square of $x - \dfrac{3}{2}$, so

$$f(x) = \left(x - \frac{3}{2}\right)^2 + \frac{7}{4}. \quad \blacksquare$$

Example 3:   Complete the square for $f(x) = 4x^2 + 20x - 100$.

Solution: Now $a = 4$, so we must first factor it out of the first two terms:

$$f(x) = 4(x^2 + 5x) - 100.$$

The coefficient of the x-term is 5, halving and squaring it gives $\dfrac{25}{4}$ as the term to add and subtract, so

$$f(x) = 4\left(x^2 + 5x + \frac{25}{4} - \frac{25}{4}\right) - 100$$

(Notice the $\dfrac{25}{4}$ is subtracted <u>inside</u> the bracket.)

$$= 4\left(x^2 + 5x + \frac{25}{4}\right) - 25 - 100$$

$$= 4\left(x + \frac{5}{2}\right)^2 - 125. \quad \blacksquare$$

Sometimes we need to complete the square in an equation.   We may also need to complete the square in more than just one variable.   Check the next example.

Example 4: Complete the square in x and y for $x^2 - 4x + y^2 + 6y = 2$.

Solution: Since half of the x-coefficient is -2, when squared equals 4, we must first add 4 to both sides to complete the square in x. Next, half of the y-coefficient is 3, when squared equals 9, so we add 9 to both sides giving

$$(x^2 - 4x + 4) + (y^2 + 6y + 9) = 2 + 4 + 9.$$

(Notice that instead of adding and subtracting on the left side, we added the same amount to both sides, which amounts to the same thing.)

So $$(x - 2)^2 + (y + 3)^2 = 15.$$

[By the way, this equation is that of a circle of radius $\sqrt{15}$ centered at the point $(2, -3)$.] ∎

The particular problem you solve will determine whether or not you choose to add the number to both sides of the equation, or just add and subtract the number on one side. The results are equivalent.

Example 5: Complete the square in x and y for $4x^2 - 9y^2 + 8x + 18y - 25 = 0$.

Solution: First regroup terms and factor out the coefficients of the quadratic terms:

$$4(x^2 + 2x) - 9(y^2 - 2y) - 25 = 0$$

Now add and subtract appropriate constants: in this case both are 1.

$$4(x^2 + 2x + 1 - 1) - 9(y^2 - 2y + 1 - 1) - 25 = 0.$$

Upon simplifying:

$$4(x + 1)^2 - 4 - 9(y - 1)^2 + 9 - 25 = 0$$

or $$4(x + 1)^2 - 9(y - 1)^2 = 20. ∎$$

Example 6: Complete the square in x and y for $x^2 - \pi x + 2y^2 - y = 0$.

Solution: For the quadratic in x, we need to add and subtract $\dfrac{\pi^2}{4}$, while for the y-terms we need to first factor out the 2 and then add and subtract $\dfrac{1}{16}$, giving

$$x^2 - \pi x + \frac{\pi^2}{4} - \frac{\pi^2}{4} + 2(y^2 - \frac{1}{2}y + \frac{1}{16} - \frac{1}{16}) = 0.$$

Simplifying gives

$$\left(x - \frac{\pi}{2}\right)^2 + 2\left(y - \frac{1}{4}\right)^2 = \frac{\pi^2}{4} + \frac{1}{8} = \frac{2\pi^2 + 1}{8}. \quad \blacksquare$$

Exercises A.1

1) Complete the square for the following expressions:

    a) $f(x) = x^2 - 6x + 15$ b) $h(y) = y^2 + 5y$

    c) $g(s) = s^2 + 2s - 8$ d) $k(x) = 2x^2 - 2x + 5$

    e) $f(x) = 3x^2 - 7x + 1$ f) $w(x) = \pi x^2 + 2x$

2) Complete the square for the following equations:

    a) $x^2 - 3x - 17 = 0$ b) $-3x^2 - 6x + 15 = 0$

3) Complete the square in both x and y for the following equations:

    a) $x^2 + 3x + 2y^2 - 8y = 0$ b) $3x^2 + 6x - 2y^2 - 8y = -11$

    c) $-x^2 + 4x + y^2 - 16y = 40$ d) $-9x^2 + 36x - 4y^2 - 8y = 0$

    e) $x^2 + y^2 - 6x + 10y + 34 = 0$

The graph of this last example is called a degenerate circle. (Can you figure out why?)

A.2 Derivation of the Quadratic Formula

In the last section we demonstrated how to complete the square for a variety of different situations involving quadratic functions. Here's an example of just how useful the method is.

Example 1: Complete the square to solve $ax^2 + bx + c = 0$.

Solution: First let us consider the algebraic expression $ax^2 + bx + c$. If we wish to complete the square for this general quadratic polynomial we first need to factor out the a from the first and second terms in the expression. That is:

$$ax^2 + bx + c = a\left(x^2 + \frac{b}{a}x\right) + c.$$

Now, we need to complete the square inside the parentheses. The coefficient of the x-term is $\frac{b}{a}$, half of that is $\frac{b}{2a}$, and squaring $\frac{b}{2a}$ gives $\frac{b}{4a^2}$. So we need to add and subtract $\frac{b}{4a^2}$ inside the parentheses. Doing so we get

$$ax^2 + bx + c = a\left(x^2 + \frac{b}{a}x + \frac{b^2}{4a^2} - \frac{b^2}{4a^2}\right) + c.$$

Removing the last term from the parentheses gives

$$ax^2 + bx + c = a\left(x^2 + \frac{b}{a}x + \frac{b^2}{4a^2}\right) - a\frac{b^2}{4a^2} + c$$

$$= a\left(x + \frac{b}{2a}\right)^2 - \frac{b^2}{4a} + c$$

$$= a\left(x + \frac{b}{2a}\right)^2 - \frac{b^2 - 4ac}{4a}.$$

If $ax^2 + bx + c = 0$, then from the last equation we have

$$a\left(x + \frac{b}{2a}\right)^2 - \frac{b^2 - 4ac}{4a} = 0.$$

This equation is equivalent to

$$a\left(x + \frac{b}{2a}\right)^2 = \frac{b^2 - 4ac}{4a}.$$

Let's "peel the onion" to solve for x. (See Section 2.1 if you're not sure how to do this.) First get rid of the coefficient a:

$$\left(x + \frac{b}{2a}\right)^2 = \frac{b^2 - 4ac}{4a^2}.$$

Take square roots of both sides:

$$x + \frac{b}{2a} = \pm\sqrt{\frac{b^2 - 4ac}{4a^2}} = \pm\frac{\sqrt{b^2 - 4ac}}{\sqrt{4a^2}}.$$

Notice that $\sqrt{4a^2}$ equals $2a$ if $a > 0$, and equals $-2a$ if $a < 0$. Hence because of the $\pm$ sign, $\pm\dfrac{\sqrt{b^2 - 4ac}}{\sqrt{4a^2}}$ equals $\pm\dfrac{\sqrt{b^2 - 4ac}}{2a}$ in either case.

So

$$x + \frac{b}{2a} = \pm\frac{\sqrt{b^2 - 4ac}}{2a}.$$

Subtracting $\dfrac{b}{2a}$ from both sides of this equation gives

$$x = -\frac{b}{2a} \pm \frac{\sqrt{b^2 - 4ac}}{2a}$$

$$\boxed{\;= \frac{-b \pm \sqrt{b^2 - 4ac}}{2a}\;}.$$

Look familiar? ■

Appendix B

Factorials

Factorials come up a lot in Probability, but not so much in Calculus, at least not until you get to infinite series. So you may not need this early in your Calculus course; just remember that it's here.

In some contexts, you will see expressions like $1 \cdot 2 \cdot 3 \cdot 4$, which is of course the number 24, right? This can also be written in the form 4!, and is read "four factorial." Similarly $5! = 1 \cdot 2 \cdot 3 \cdot 4 \cdot 5 = (4!) \cdot 5$, and since $4! = 24$, $5! = 24 \cdot 5 = 120$. We can continue this: $6! = (5!) \cdot 6 = 720$ and $7! = (6!) \cdot 7 = 5040$. You can see that these little things called factorials get pretty large pretty quickly.

Definition: Let n be a positive whole number. We define $n!$ to be $1 \cdot 2 \cdot 3 \cdot 4 \cdots (n-1) \cdot n$, and call it "$\underline{n}$ factorial." We also define 0! to be 1. (We can define 0! to be anything we like; it turns out that this definition makes the theorems of probability and infinite series most elegant. When it comes to theorems, mathematicians love elegance.)

As you may have figured out by now $n! = n \cdot (n-1)!$. This allows for interesting mathematics. The fun starts when you take quotients of factorials.

> **Example 1:** Express the following as whole numbers.
>
> a) 8!   b) $\dfrac{10!}{9!}$ c) $\dfrac{1000!}{998!}$
>
> **Solution:** a) $8! = (7!) \cdot 8 = 5040 \cdot 8 = 40320$
>
> b) For this one, we could calculate the top and bottom separately, and then divide, but what a drudge that would be! But check this:
>
> $$\frac{10!}{9!} = \frac{(9!) \cdot 10}{9!} = 10. \text{ We're done.}$$
>
> c) $\dfrac{1000!}{998!} = \dfrac{1000 \cdot 999 \cdot (998!)}{998!}$
> $= 1000 \cdot 999 = 999{,}000.$ ■

Exercises B.1

1)   Simplify the following expressions:

    a) $\dfrac{12!}{11!}$ b) $\dfrac{6!}{8!}$ c) $\dfrac{7!}{5!\cdot 2!}$

2)   Compute $\dfrac{n!}{k!(n-k)!}$ for $n=6$ and $k=2$.

3)   Compute $\dfrac{n!}{k!(n-k)!}$ for $n=6$ and $k=1$.

4)   Compute $\dfrac{n!}{k!(n-k)!}$ for $n=6$ and $k=0$.

5)   In probability, there is a symbol $_nC_r$ defined as $\dfrac{n!}{r!(n-r)!}$. What can you say about $_nC_r$ and $_nC_{n-r}$? (Hint: Play around. Try $n=8$ and $r=3$, for example.)

Appendix C

The Binomial Theorem

A binomial is the sum of two terms and can therefore be represented as $a + b$. The binomial theorem states what happens when powers of $a + b$ are multiplied out. Let's expand $(a + b)^n$ for $n = 1, 2, 3, 4$.

$$(a + b)^1 = a + b$$

$$(a + b)^2 = a^2 + 2ab + b^2$$

$$(a + b)^3 = a^3 + 3a^2 b + 3ab^2 + b^3$$

$$(a + b)^4 = a^4 + 4a^3 b + 6a^2 b^2 + 4ab^3 + b^4$$

We see some obvious patterns. The powers of a decrease as you go from left to right, starting with a^n and ending with a^0 (which equals 1, so it doesn't appear). The powers of b go in reverse, from $b^0 = 1$ to b^n. But what about the coefficients? Let's write them in the form known as Pascal's triangle:

$$
\begin{array}{ccccccc}
 & & & 1 & 1 & & \\
 & & 1 & 2 & 1 & & \\
 & 1 & 3 & 3 & 1 & & \\
1 & 4 & 6 & 4 & 1 & &
\end{array}
$$

Each line begins with 1, followed by the line number, n. Also we see that each entry of a particular line is formed by adding the two entries diagonally above it. For example, in the fourth line, 4 is the sum of $1 + 3$, 6 is the sum of $3 + 3$, etc. To get the next line, notice that $1 + 4 = 5$, $4 + 6 = 10$, etc., – which means that the next line is 1 5 10 10 5 1.

Binomial Theorem: Let n be a positive whole number, and let a and b be any numbers. Then

$$(a + b)^n = (\ ) a^n + (\ )a^{n-1} b + (\ )a^{n-2} b^2 + \cdots + (\ )a^2 b^{n-2} + (\ )ab^{n-1} + (\ )b^n,$$

where the coefficients in the empty brackets are given by the entries of the nth line of Pascal's triangle.

Example 1:   Expand $(a + b)^6$.

Solution: Check Pascal's triangle. The fifth and sixth lines are:

$$1 \quad 5 \quad 10 \quad 10 \quad 5 \quad 1$$

$$1 \quad 6 \quad 15 \quad 20 \quad 15 \quad 6 \quad 1$$

Hence $(a + b)^6$ is equal to

$$a^6 + 6a^5 b + 15a^4 b^2 + 20a^3 b^3 + 15a^2 b^4 + 6ab^5 + b^6. \quad \blacksquare$$

Example 2:   Expand $(x + 2)^5$.

Solution: Check Pascal's triangle. The fifth line is:

$$1 \quad 5 \quad 10 \quad 10 \quad 5 \quad 1$$

Hence $(x + 2)^5$ is equal to

$$(1)x^5 + (5)x^4 2 + (10)x^3 2^2 + (10)x^2 2^3 + (5)x 2^4 + (1)2^5,$$

which can be simplified to give

$$(x + 2)^5 = x^5 + 10x^4 + 40x^3 + 80x^2 + 80x + 32. \quad \blacksquare$$

Example 3:   Expand $(a - 3)^4$.

Solution: $(a - 3)^4 = (a +(-3))^4$ (Tricky!)

$$= a^4 + (4)a^3 (-3) + (6)a^2 (-3)^2 + (4)a(-3)^3 + (1)(-3)^4$$

$$= a^4 - 12a^3 + 54a^2 - 108a + 8 . \blacksquare$$

Remarks: a) This method is clearly simpler than multiplying $a - 3$ by itself repeatedly, but it can become cumbersome if the number n is too big.

b) The coefficients, which appear in Pascal's triangle, can also be expressed by a formula. This can be found in calculus books in the section on binomial series.

Example 4:   Simplify $\dfrac{(x+h)^4 - x^4}{h}$.

Solution: Check Pascal's triangle. The fourth line is:

$$1 \quad 4 \quad 6 \quad 4 \quad 1$$

Hence

$$(x+h)^4 = (1)x^4 + (4)x^3 h + (6)x^2 h^2 + (4)x h^3 + (1)h^4$$

$$= x^4 + 4x^3 h + 6x^2 h^2 + 4x h^3 + h^4,$$

and $\dfrac{(x+h)^4 - x^4}{h} = \dfrac{4x^3 h + 6x^2 h^2 + 4x h^3 + h^4}{h}$

$$= 4x^3 + 6x^2 h + 4x h^2 + h^3. \quad \blacksquare$$

Remark: If we expand $(x+h)^n$, we get $x^n + nx^{n-1}h + (\ )h^2 + (\ )h^3 + \cdots + h^n$,   where the open brackets contain polynomials <u>in x only</u>, without any factors of h. This fact will be crucial when finding the derivative of x^n.

Exercises C.1

1) Expand $(x+1)^6$.

2) Expand $(y-1)^6$.

3) Expand $(2z+3)^4$.

4) Expand $(x+\Delta x)^5$.

5) Using the result in Exercise 4, simplify $\dfrac{(x+\Delta x)^5 - x^5}{\Delta x}$.

Answers to Exercises

Chapter 1

1.1 page 3

1) -19 2) 46 3) $4xy - x + 2y$

4) 0 5) $3xy - 6x - xy^2 + 2y$ 6) $2xz - 3yz$

7) $x^2y + xy^3 - 4x^2y^2 + 2y^3$

8) $x^3y^2 - xy^4 - 8x^2y^2 + 6x^2y^3 + 2xy^2$

9) a) 27 in 8 operations b) 27 in 11 operations, so method 1 is "cheaper."

1.2 page 7

1) $\frac{1}{4}$ 2) $\frac{\pi^2}{2}$ 3) $\frac{3}{5}$ 4) $\frac{1}{6}$ 5) $\frac{-28}{51}$ 6) $\frac{9}{14}$

7) $\frac{21y + 14}{3y}$ 8) $\frac{x - xy + 2 - 2y}{(1 + y)x}$ 9) $\frac{y}{y - 2}$ 10) $\frac{w}{xy}$

11) $\frac{(x + y)^2}{x}$ 12) $\frac{1}{y(x - y)}$

13)

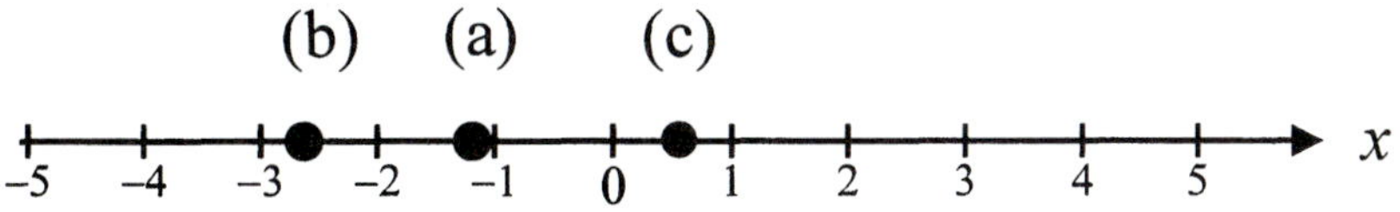
(b)
(a)
(c)
−5
−4
−3
−2
−1
0
1
2
3
4
5
x

14) $\frac{-5}{1}, \frac{8}{1}, \frac{3857}{1000}$

1.3 page 10

1) $\frac{7}{12}$ 2) $\frac{11}{8}$ 3) $\frac{7}{30}$ 4) $\frac{5}{16}$ 5) $\frac{37}{30}$

6) $\frac{-16}{9}$ 7) $\frac{22}{45}$ 8) $\frac{-25}{66}$ 9) $\frac{29}{42}$ 10) $\frac{x+y}{xy}$

11) $\frac{x-y}{xy}$ 12) $\frac{4yz - 2xz + xy}{xyz}$ 13) $\frac{yz - z(x + 1) + y(x - 2)}{xyz}$

14) $\frac{x(z - xy)}{y(x - z)}$ 15) $\frac{w - st}{s - 2tw}$ 16) $\frac{y^2 - x^2}{xy}$

17) $\frac{4y^3z^2 - 2xz^2 + xy}{x^2y^2z}$ 18) $\frac{xy^2z - x^3yz + 2x^3 + 2xy^2 - 2x^2y - 2y^3}{2x^3yz + 2xy^3z}$

1.4 page 13

1) $\dfrac{225}{16}$ 2) $\dfrac{9}{2}$ 3) $\dfrac{60}{19}$ 4) $\dfrac{4}{81}$ 5) $32\frac{1}{2}$

6) yz^8 7) $\dfrac{1}{4}$ 8) x^{68} 9) x^{-68}

10) Try $x = y = 1$. L.H.S. $= 2^2$, while R.H.S. $= 2$.

11) x 12) $y^{-12} = \dfrac{1}{y^{12}}$ 13) $\dfrac{x^2}{y}$ 14) $\dfrac{x^6 z - x^2}{y^5 z^5}$ 15) $\dfrac{x\,y}{x + y}$

1.5 page 15

1) 12 2) −4 3) $\dfrac{1}{3}$ 4) −2 5) $\dfrac{2}{7}$

6) $\dfrac{2}{3}$ 7) 32 8) −32 9) 4 10) −4

11) $\dfrac{27}{64}$ 12) 1000 13) .0016 14) $2^{3/15}$ 15) 256

16) $3^{1/2}$ 17) $2^{-13/14}$ 18) $3^6 = 729$ 19) $2^{-1/15}$ 20) $3 + 2 = 5$

21) Yes! $\sqrt{a^2 b^2} = \sqrt{a^2}\,\sqrt{b^2} = |a| \cdot |b|$ But, if $a > 0$ and $b > 0$ $|a| = a$ and $|b| = b$, so
$\sqrt{a^2 b^2} = |a| \cdot |b| = a\,b$.

22) No! Let $a = b = 1$, $\sqrt{2}$ is certainly not equal to 2.

23) No! Let $x = 1$, $\sqrt[3]{-7}$ is not equal to -1.

24) No! Let $x = y = \dfrac{5}{\sqrt{2}}$, then $x^2 + y^2 = 25$ but $x + y = \dfrac{10}{\sqrt{2}}$.

25) No! Let $a = b = 1$, $\sqrt{2}$ is certainly not equal to 2.

1.6 page 18

1) 50 2) .2343 3) 1304.444 . . . 4) a) $12.75 b) $1124.25
5) $34.00 6) $191.20 7) $240.00 8) $52,500.00
9) a) $8000 b) $11,000 c) $3000 d) 37.5%

1.7 page 21

1) a) 3.83×10^5 b) -7.24×10^{-4} c) 3.00 d) 2.00×10^2
2) a) 9.48×10^7 b) -3.09×10^5 c) 3.50×10^{-100} d) 7.66×10^{-2} e) -1.68×10^{-8}
   f) 5.57×10^{-2}
3) a) $8017.84 b) $3714.87 48.6% increase
4) a) 1.83×10^9 b) 6.22×10^{10} c) 34 d) 3299%

1.8 page 23

1) a) $[-1,3]$

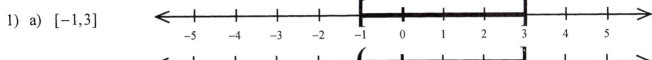

   b) $(-1,3]$

   c) $[-3,1)$

   d) $[-3,4]$

   e) $(-\frac{1}{2},\sqrt{2}\,]$

   f) $[\pi,5]$

   g) $(0,\infty)$

   h) $[3,\infty)$

   i) $(-\infty,-4)$

   j) $[3-\pi,\infty)$

   k) $(-\infty,5)$

   l) $(-\infty,3]$

2) a) $3<x<7$ b) $-4<x\le-1$ c) $x\le19$ d) $2\le x<10$ e) $-2\le x\le-1$

3) a) $[3,5)$ b) $(-\infty,\infty)$ c) $\varnothing$ Empty Set: d) $[4,7]$

4) $(-7,1]$

   $(3,7)$

   $(-7,1]\cup(3,7)$
   No you can't.

Chapter 2

2.1 page 31

1) $\dfrac{4}{3}$ 2) 84   3) $\dfrac{33}{40}$ 4) $\dfrac{75}{7}$ 5) $\dfrac{1}{5}$ 6) $\dfrac{-z^2}{3y^2}$

7) $-\dfrac{3z+1}{2z}$ 8) $-\dfrac{1}{2y+3}$ 9) $\dfrac{y^2+3y+1}{2y^2-1}$ 10) $\dfrac{z+y^2}{2zy^2+2y+1-z^2}$

2.2 page 37

1) $-1,-4$ 2) $-2,4$ 3) repeated root, -3 4) $\dfrac{1}{2},-1$

5) Repeated root; 4   6) $-4\pm2\sqrt{3}$ 7) $1\pm i$

8) $\pm\dfrac{2}{\sqrt{3}}=\pm\dfrac{2\sqrt{3}}{3}$ 9) $1,-4$ 10) $\dfrac{-7\pm\sqrt{33}}{2}$

11) $-1,5$ 12) $-\dfrac{1}{2},-\dfrac{1}{4}$ 13) $\dfrac{-7\pm\sqrt{17}}{4}$ 14) $\dfrac{1}{3},-\dfrac{1}{5}$

15) $-3\pm2\sqrt{3}$ 16) $\pm\sqrt{3}\,i$ 17) $-1\pm\sqrt{5}$ 18) $-0.4,1$

19) $-y\pm\sqrt{3}\,y=-(1+\sqrt{3})y$, and $-(1-\sqrt{3})y$ 20) $2(y+z)\pm2\sqrt{2yz}$

21) $\dfrac{x}{2}\pm\sqrt{xy-y^2}$

2.3 page 44

1) $3,-6$ 2) $-2,9$ 3) $-\dfrac{1}{2},2$ 4) $3,-3,-6$

5) $\pm\sqrt{5},\pm4$ 6) $\pm i,12$ 7) $-5,1,\pm3\sqrt{3}$ 8) ±2, each root is repeated

9) $\pm2\sqrt{2},\pm2\sqrt{2}\,i$ 10) $\pm2,\pm\sqrt{5}$

11) 16   12) 16   13) 2   14) $\frac{1}{4}$ or $\frac{1}{16}$

15) 1   16) 4   17) 1, 32   18) 1

19) No solution   20) 2   21) $-1,3$

Chapter 3

3.1 page 50

1) a) 540   b) 580 on May 6th   c) 320 on May 19th   d) 487

e)

Week of April 19	Volume
Monday	380
Tuesday	518
Wednesday	480
Thursday	480
Friday	400

2) a)1500 thousand   b) The sheep population has been constant since 1864.

c)

Year	Population
1814	60
1824	260
1834	875
1844	1500
1854	1625
1864	1625

3) a)

b) $C = 6 + 1.5n$

3.2 page 56

1) a) The cost of a CD as a function of how many CD's you buy
    b) The domain consists of the set of positive integers
        The range is the set { 5.99, 6.49, 6.99, 7.99 }
    c) $ 58.41

2) a) Car thefts-per-year as a function of year.
    b) Domain is the set of years { 1990, 1991, 1992, 1993, 1994, 1995, 1996 }
        Range is the set of the number of car thefts { 33, 34, 38, 42, 45, 55, 67 }

3) a) 12   b)   33   c) $(x + h)^3 + 2(x + h)$ · d) $8x^3 + 4x$
    e) $-x^3 - 2x$ f) $(2 + \Delta x)^3 + 4 + 2\Delta x$

4) a) 1   b) 1   c)   16   d) $\dfrac{11}{15}$

5) a) 0   b) 254   c) $(x + h)^4 - \sqrt{x + h}$ d) $\dfrac{\pi^8}{256} - \dfrac{\pi}{2}$ e) $x^4 - \sqrt{-x}$

6) All x such that $-2 < x < 2$. See Exercise 3.5 # 5.

7) All x that are less than −3 or greater than 3, i.e., the solution set is $(-\infty, -3) \cup (3, \infty)$.

3.3 page 65

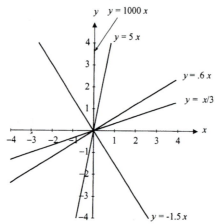

Believe it or not the graph for $y = 1000\,x$ is really plotted on this graph, but because of the scale you can't really see it. If we scale this as in the next figure you see this but lose the others.

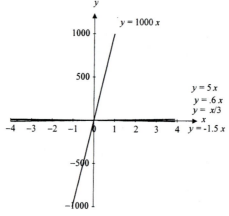

6) $y - 7 = \dfrac{2}{3}(x - 2)$ 7) $y - 2 = -6(x + 3)$

8) $m = \dfrac{2 - 1}{1 - (-2)} = \dfrac{1}{3}$, so $y - 2 = \dfrac{1}{3}(x - 1)$ 9) $(y - 4) = -(x + 1)$

10) The x-intercept is $(-8, 0)$, y-intercept is $(0, 6)$

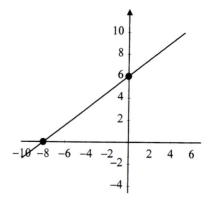

11) $y - 4 = 2(x - 1)$ or $y = 2x + 2$

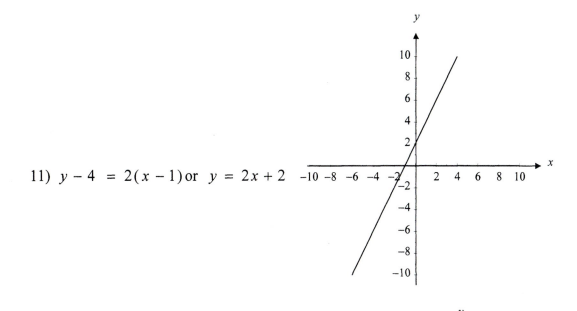

12) $y - 4 = -3(x + 1)$ or $y = -3x + 1$

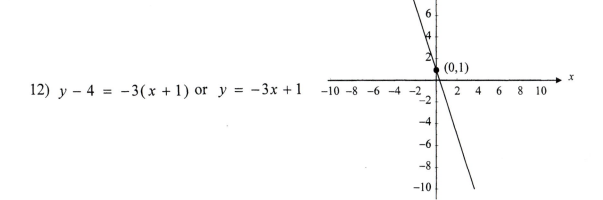

13) $y = 5x + \dfrac{4}{3}$

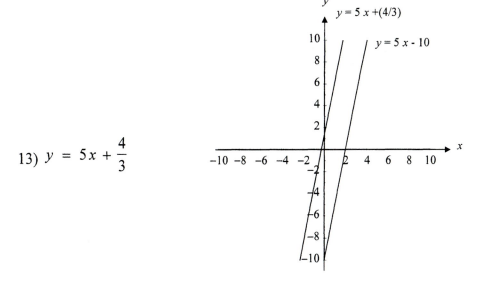

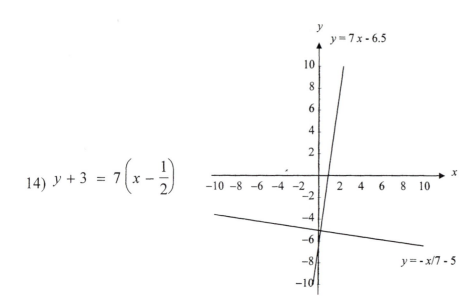

14) $y + 3 = 7\left(x - \dfrac{1}{2}\right)$

3.4 page 70

1) We list four points, given in the table and plotted on the graph. By picking more and more points, we can "fill in" the picture, getting the solid line in the figure below.

x	$\sqrt{x}$
0	0
1	1
4	2
9	3

Exercises 2–6 and Exercise 8 are all shown in the section body.

7) First we notice that the function $x^{2/3}$ is defined for both positive and negative values of x. Some values are given in the following table, and the graph is shown in the following figure.

x	$x^{2/3}$
0	0
1	1
−1	1
8	4

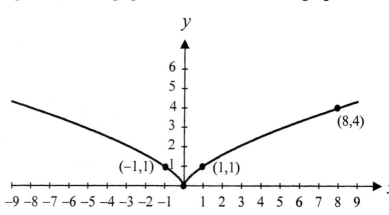

3.5 page 72

1)

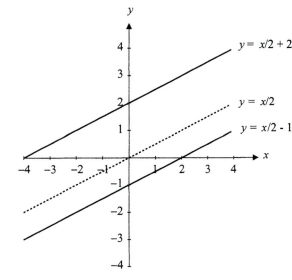

2a),b)

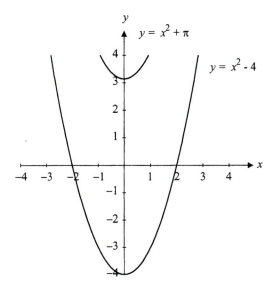

2c)

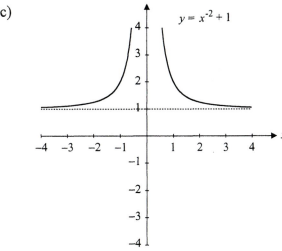

3)

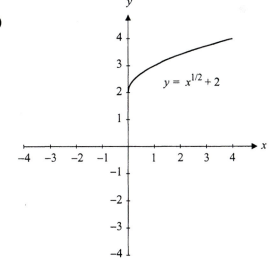

4)

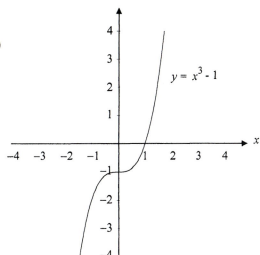

5)

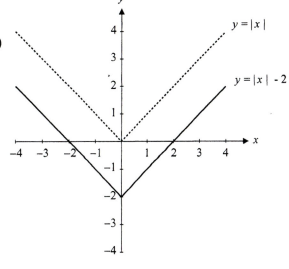

3.6 page 74

1)

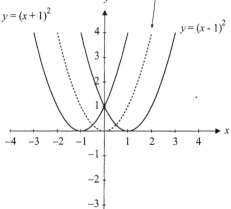

2a),b)

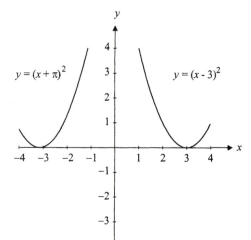

2c)

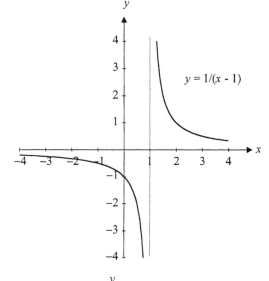

3a)

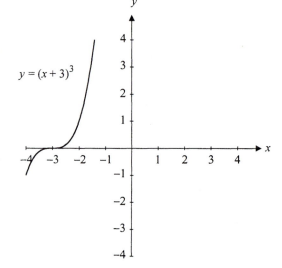

3b)

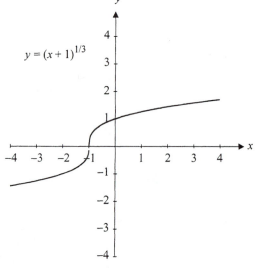

3c)

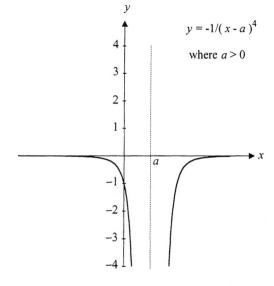

4)

$$y = (x - 4)^{1/2}$$

5) $g(x) = f(x+3) - 2$

6) $g(x) = f(x-2) + 1$ 7) $g(x) = f(x-h) + k$

8) $g(x) = -f(x) = -x^2, \quad k(x) = g(x-2) - 1 = -(x-2)^2 - 1$

3.7 page 78

1) a) $x = 2$, $y = -5$ b) $x = 2$, $y = -\dfrac{1}{2}$ and $x = -6$, $y = \dfrac{15}{2}$

2) a) Curves do not intersect.   b) $x = 4$, $y = -2$

Chapter 4

4.1 page 79

1) $2x(y + 2)$ 2) $2wz(3 + t)$ 3) $x(y + 4 + 2w)$

4) $3xy(2x + 1 + 3y)$ 5) $5x^2 y^4 (2x^6 y^2 + 5 + 4xy^6)$

6) $2xyz(12x + yz + 2z^2)$

4.2 page 83

1) $(2y + 3z)(2y - 3z)$ 2) $(4x^2 - y^3)(4x^2 + y^3)$

3) $(2s + 3t)(4s^2 - 6st + 9t^2)$ 4) $(2^{\frac{1}{3}} x + 4y)(2^{\frac{2}{3}} x^2 - 4 \cdot 2^{\frac{1}{3}} xy + 16y^2)$

5) $(2s - 3t)(4s^2 + 6st + 9t^2)$ 6) $(4z - 9^{\frac{1}{3}} t)(16z^2 + 4 \cdot 9^{\frac{1}{3}} zt + 9^{\frac{2}{3}} t^2)$

7) $(x + 1)^2$ 8) $(x + 4)(x + 2)$ 9) $(x - 6)(x + 4)$

10) $(a - \sqrt{6})(a + \sqrt{6})(a^2 + 4)$ 11) $(s - 2)(s^2 + 2s + 4)(s + 1)(s^2 - s + 1)$

12) $(3x - 2)(x + 1)$

4.3 page 85

1) $(3x + 2y)(a + b)$ 2) $(x - y)(x + 1)(x^2 - x + 1)$

3) $(x^4 + y^2)(x^2 + y)(x^4 - x^2 y + y^2)$ 4) $(\sqrt{3} x + \sqrt{2} y)(\sqrt{3} x - \sqrt{2} y)(2y)(x + 2)$

5) $(x + y)(3x + 5y + 7)$

4.4 page 89

1) $(x-1)^2(x+2)$

2) $(x+1)(2x^2-2x+3)$

3) $(x+2)(x^2-2x+3)$

4) $2x^2-3x+4$

5) $\left(x-\dfrac{3+\sqrt{17}}{2}\right)\left(x-\dfrac{3-\sqrt{17}}{2}\right)$

6) $24(x-3)(x+1)$

4.5 page 92

1) $7(\sqrt{2}+1)$

2) $\dfrac{x^2-2}{2(x-\sqrt{2})}$

3) $\dfrac{3(x+\sqrt{7})}{x^2-7}$

4) $\dfrac{(x+1)(x-\sqrt{11})}{x^2-11}$

5) $x+\sqrt{3}$

6) $(x^2+6)(x-\sqrt{6})$

7) $(x^4+3)(x^2-\sqrt{3})$

8) $\dfrac{-2}{\sqrt{2(x+h)}\,\sqrt{2x}\left(\sqrt{2x}+\sqrt{2(x+h)}\right)}$

4.6 page 95

1) $4|x|$

2) $2|x|\sqrt{1+2x^2}$

3) $3x\sqrt[3]{2x}$

4) $x^6\sqrt{3y}$

5) $x^2\sqrt{5+3x^4}$

6) $3x^2\sqrt[3]{y}$

7) $2\pi y^2 x\sqrt{2x}$, for $x\ge 0$

8) $|xy|\sqrt[4]{x+x^2y^6}$

Chapter 5

5.1 page 103

1) a) $(x+h)^2+3(x+h)$

    b) $\dfrac{\left((x^2+2xh+h^2)+3x+3h\right)-(x^2+3x)}{h}=2x+h+3$

2) a) $2(x+h)^3-(x+h)$ b) $6x^2+6xh+2h^2-1$

3) $\dfrac{1}{x(x-1)}$

4) $\dfrac{x^2-2x+2}{x(x-1)}$

5) $\dfrac{2x}{(x^2-1)}$

6) $2(s^2+1)$

7) 2

8) $\dfrac{-2x-h}{x^2(x+h)^2}$

9) $\dfrac{x^4+3x^2+1}{x^3+2x}$

10) $\dfrac{2}{\sqrt{2x+2h}+\sqrt{2x}}$

11) $\dfrac{a-9b}{75a^5b^7}$

12) $\dfrac{1}{\sqrt{x+\Delta x-3}+\sqrt{x-3}}$

Chapter 6

6.1 page 108

1) $x + 1$

2) $\dfrac{1}{x^3 + 2x + 2}$

3) a) $(x + 1)^{3/2}$ b) $(2t - 3)^3$ c) $\left(x^2 - \dfrac{1}{x}\right)^3$ d) $\sqrt{x^3 + 1}$ e) $\sqrt{2t - 2}$

f) $\sqrt{x^2 - \dfrac{1}{x} + 1}$ g) $2x^3 - 3$ h) $2\sqrt{x + 1} - 3$ i) $2\left(x^2 - \dfrac{1}{x}\right) - 3$

j) $x^6 - \dfrac{1}{x^3}$ k) $x + 1 - \dfrac{1}{\sqrt{x + 1}}$ l) $(2t - 3)^2 - \dfrac{1}{2t - 3}$

4) a) $\dfrac{1}{x + 1} - \dfrac{2}{\sqrt{x + 1}}$ b) $\left(\dfrac{1}{\sqrt{x} + 1}\right)^2 - \dfrac{2}{\sqrt{x} + 1}$

c) $\dfrac{1}{|x - 1|}$ d) $\sqrt{\dfrac{1}{(x + 1)^2} - \dfrac{2}{x + 1}}$

6.2 page 112

1) a) Outer function $f(x) = \sqrt{x}$, inner function $g(x) = x + 1$.
   b) Outer function $f(x) = x^2$, inner function $g(x) = x^3 - 1$.
   c) Outer function $f(x) = x^7$, inner function $g(x) = x^{7/2} + x$.
   d) Outer function $f(x) = \sqrt[3]{x}$, inner function $g(x) = x^{2/3} + 2$.
   e) Outer function $f(x) = x^{-3/2}$, inner function $g(x) = x^5 + 3x^2 + x$.

2) a) $f(x) = \ln x$, $g(x) = x^2 + 2$ b) $f(x) = e^x$, $g(x) = \sqrt{x} + 1$
   c) $f(x) = x^2$, $g(x) = \ln x + e^x$ d) $f(x) = \ln x$, $g(x) = x + e^x$
   e) $f(x) = \sqrt{x}$, $g(x) = \ln\left(x^2 + 1\right)$

Chapter 7

7.1 page 116

1) -1 2) $\dfrac{5}{3}$ 3) $\dfrac{-x}{3y}$ 4) $\dfrac{x^2 + 6x - 1}{x(x + y)}$

5) $\dfrac{-y(2 + 3xy^2)}{x(1 + 3xy^2)}$ 6) $\dfrac{x(3 - y - 2y^2)}{2x^2y + x + 2y}$ 7) $\dfrac{x(1 + 2y)}{y^2 - 4x^2}$

Chapter 8

8.2 page 126

1) Let L = the length of the fencing, and x = the width of the field. Then $L = 2x + \dfrac{20{,}000}{x}$ ft.

2) Let P = the perimeter of the window, A = the area of the window, and x = the width of the window. Then $P = 5x + \dfrac{\pi}{2}x$, and $A = x^2\left(2 + \frac{\pi}{8}\right)$ ft.

3) Let x = width a) $A = 2x^2$ b) $P = 6x$ c) $A = 2\left(\dfrac{P}{6}\right)^2 = \dfrac{1}{18}\cdot P^2$ d) $P = \sqrt{18}\cdot\sqrt{A}$

4) Let r = the radius of the can, and h = the height of the can, then $h = \dfrac{100 - 2\pi r^2}{2\pi r}$ in.

5) Let V = the volume of the box, x = the length of the square cutout, then $V = 4x^3 - 36x^2 + 80x$ in³.

6) With variables defined as in Exercise 5) we have $V = 4x^3 - 7x^2 + 3x$ in³. If A = the exterior surface area then Exterior surface area = $A = (1.5)(2) - 4x^2 = 3 - 4x^2$.

7) $V = \dfrac{S^{3/2}}{6\sqrt{\pi}}$ and $S = \sqrt[3]{36\pi}\ V^{2/3}$.   8) $A = \dfrac{x^2}{4\pi} + \dfrac{(2-x)^2}{18}$

9) golden ratio $= \dfrac{1 + \sqrt{5}}{2}$

10) a) $50P$ b) $(51)(.99P) = 50.49P$ c) $(80)(.7P) = 56P$ d) $(x)\left(1 - \dfrac{x-50}{100}\right)P$

Chapter 9

9.1 page 132

1)

As $x \to \infty$, $.32^x$ and $(\frac{2}{3})^x$ both approach 0, while $(\frac{3}{2})^x$ and 1.1^x both approach ∞.

As $x \to -\infty$, $.32^x$ and $(\frac{2}{3})^x$ both approach ∞, while $(\frac{3}{2})^x$ and 1.1^x both approach 0.

2)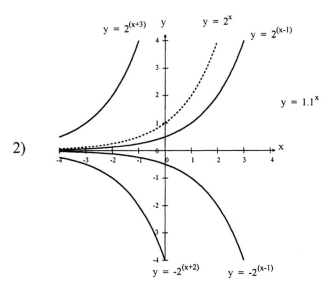

3)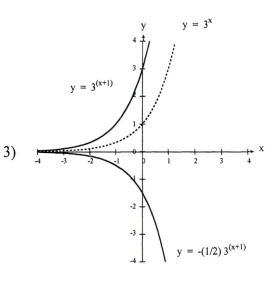

9.2 page 133

1)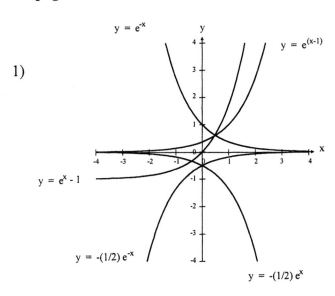

2)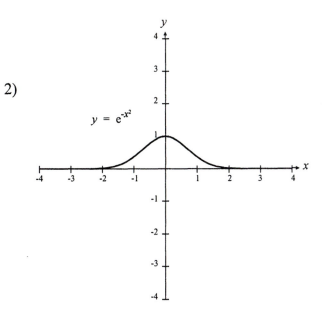

9.4 page 140

1)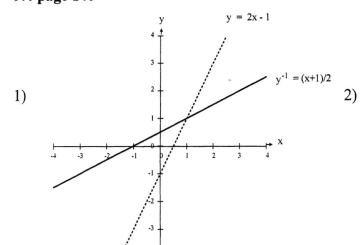

2)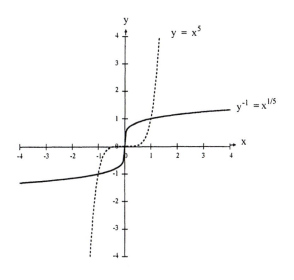

3) The function $g(x)$ is its own inverse.

4) Inverse does not exist because $y^4 + 1$ flunks the horizontal line test on the real line.

5)

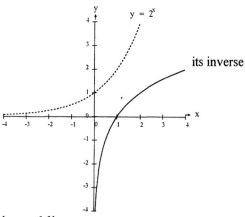

6) Also flunks the horizontal line test.

7)

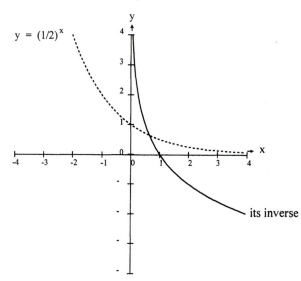

8)

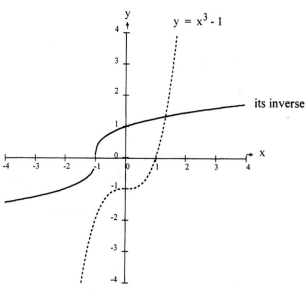

9) On $[-1,1]$ $f(x) = 1 - x^2$ has no inverse, but on $[0,1]$ an inverse does exist. (Check by graphing $1 - x^2$.)

9.5 page 143

1) $f^{-1}(x) = \dfrac{x+3}{2}$ 2) $k^{-1}(x) = \dfrac{x}{1-x}$ 3) $g^{-1}(x) = \dfrac{x^3-1}{5}$

4) $s^{-1}(t) = t^2 - 2$ 5) $f^{-1}(x) = \dfrac{2}{x}$ 6) $w = \pm\sqrt{\dfrac{v}{1-v}}$, no inverse!

7) Flunks the horizontal line test. No inverse. If $x \geq 1$ it passes the test.

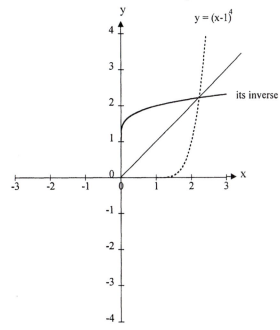

8a) Domain is $[0,1]$, range is $[3,4]$. The inverse is the function $\sqrt{x-3} = (x-3)^{1/2}$.

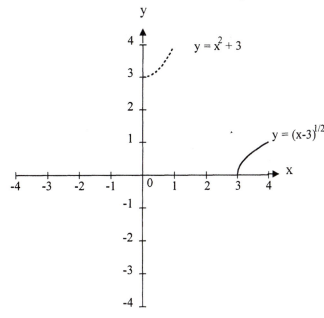

8b) Domain is [-1,0], range is [3,4]. The inverse is the function $-\sqrt{x-3} = -(x-3)^{1/2}$.

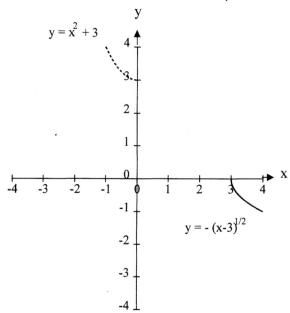

8c) No, their inverses are not the same, hence domains are very important when discussing inverse functions.

9.6 page 145

1) 2   2) -1 3) -3   4) -3   5) $\dfrac{1}{2}$ 6) $\dfrac{3}{4}$ 7) 13   8) x

9.7 page 150

1)   2)

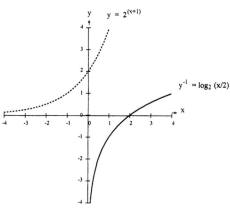

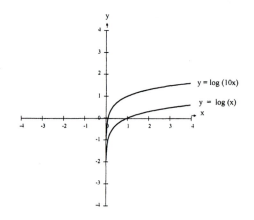

I notice that $\log 10x = 1 + \log x$.

3) a) 9   b) 2   c) $\dfrac{5}{2}$ d) -8

4) $\dfrac{-20}{3}$ 5) ± 2 6) $\pm\sqrt{14}$

9.8 page 153

1) 9   2) $\left(\dfrac{1}{10}\right)^{\frac{1}{3}}$ 3) $\dfrac{7}{10}$ 4) 4,8

5) Let $a = 2, x = y = 1$. R.H.S. $= 0$, but L.H.S. $= 1$

6) 1.26

9.9 page 157

1)   2)

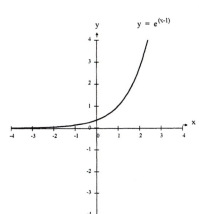

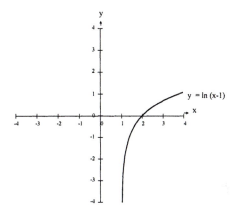

3)

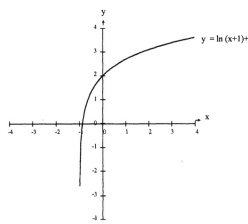

4) e^2 5)   1, -3   6) ± 2 7) $e^{2.3x}, \; e^{-.69x}$

Appendix A

A.1 page 162

1) a) $f(x) = (x-3)^2 + 6$ b) $h(y) = \left(y + \dfrac{5}{2}\right)^2 - \dfrac{25}{4}$ c) $g(s) = (s+1)^2 - 9$

  d) $k(x) = 2\left(x - \dfrac{1}{2}\right)^2 + \dfrac{9}{2}$ e) $f(x) = 3\left(x - \dfrac{7}{6}\right)^2 - \dfrac{37}{12}$ f) $w(x) = \pi\left(x + \dfrac{1}{\pi}\right)^2 - \dfrac{1}{\pi}$

2) a) $\left(x - \dfrac{3}{2}\right)^2 - \dfrac{77}{4} = 0$ b) $-3(x+1)^2 + 18 = 0$; alternatively, $(x+1)^2 - 6 = 0$

3) a) $\left(x+\dfrac{3}{2}\right)^{2} + 2(y-2)^{2} = \dfrac{41}{4}$ b) $3(x+1)^{2} - 2(y+2)^{2} = -16$

c) $-(x-2)^{2} + (y-8)^{2} = 100$ d) $9(x-2)^{2} + 4(y+1)^{2} = 40$

e) $(x-3)^{2} + (y+5)^{2} = 0$

Appendix B

B.1 page 166

1) a) 12   b) $\dfrac{1}{56}$ c) 21

2) 15   3) 6   4) 1   5) They are equal.

Appendix C

C.1 page 169

1) $x^{6} + 6x^{5} + 15x^{4} + 20x^{3} + 15x^{2} + 6x + 1$

2) $y^{6} - 6y^{5} + 15y^{4} - 20y^{3} + 15y^{2} - 6y + 1$

3) $16z^{4} + 96z^{3} + 216z^{2} + 216z + 81$

4) $x^{5} + 5x^{4}\Delta x + 10x^{3}(\Delta x)^{2} + 10x^{2}(\Delta x)^{3} + 5x(\Delta x)^{4} + (\Delta x)^{5}$

5) $5x^{4} + 10x^{3}(\Delta x) + 10x^{2}(\Delta x)^{2} + 5x(\Delta x)^{3} + (\Delta x)^{4}$

Index